I0131750

Lost Circulation: A New Approach to An Old Challenge

Society of Petroleum Engineers

Richardson, Texas, USA

Copyright © 2021 Society of Petroleum Engineers

All rights reserved. No portion of this book may be reproduced in any form or by any means, including electronic storage and retrieval systems, except by explicit, prior written permission of the publisher except for brief passages excerpted for review and critical purposes.

Printed in the United States of America.

Disclaimer

This book was prepared by members of the Society of Petroleum Engineers and their well-qualified colleagues from material published in the recognized technical literature and from their own individual experience and expertise. While the material presented is believed to be based on sound technical knowledge, neither the Society of Petroleum Engineers nor any of the authors or editors herein provide a warranty either expressed or implied in its application. Correspondingly, the discussion of materials, methods, or techniques that may be covered by letters patents implies no freedom to use such materials, methods, or techniques without permission through appropriate licensing. Nothing described within this book should be construed to lessen the need to apply sound engineering judgment nor to carefully apply accepted engineering practices in the design, implementation, or application of the techniques described herein.

ISBN: 978-1-61399-861-8 (Print)
ISBN: 978-1-61399-862-5 (ADE)

10 9 8 7 6 5 4 3 2 1

Society of Petroleum Engineers
222 Palisades Creek Drive
Richardson, TX 75080-2040 USA

http://store.spe.org
service@spe.org
1.972.952.9393

Table of Contents

Preface

Lost circulation has been a persistent challenge in the oil and gas industry since its beginning. It happens during well construction activities when drilling or completion fluids are lost to subsurface formations. Loss of wellbore fluids to the formation implies loss of control of downhole wellbore pressure, jeopardizing well integrity and putting the safety of rig personnel at risk.

Many rule-of-thumb procedures to tackle lost circulation have been proposed and implemented over the years. Because of the almost infinite variability of subsurface formations, uncertainties in downhole conditions and lack of appropriate technologies, no "one-rule-fits-all" solution has been successful. Lost circulation was not seen as a multidisciplinary task that could be tackled by acquiring and interpreting comprehensive information from different subsurface disciplines.

Recent advances, however, in geomechanics, drilling techniques, drilling fluids, cementing and lost circulation fluids, logging while drilling and data analytics, are now assisting the industry to resolve lost circulation challenges more effectively. This brief discusses the present understanding of lost circulation and cites examples of how these new techniques are being applied to resolve those challenges.

The authors would like to thank those that peer-reviewed our book and provided us with valuable feedback prior to publication.

About the Authors

Salim Taoutaou is the founder of Taoutaou Consulting LLC and currently serves as its managing director. Prior to this, he was the cementing and well integrity technical adviser for Schlumberger in Paris, where he managed the global development of technologies and solutions for oil and gas sector. Throughout his 24 years in the oil and gas industry, Taoutaou has held various managerial and technical global positions, as well as regional ones in North Africa, the North Sea, the Middle East, and Asia Pacific. He has authored more than 58 international journal and conference papers and holds several patents. In 2014, Taoutaou received the SPE Asia Pacific Regional Technical Award in Drilling Engineering. He also served as SPE Distinguished Lecturer for the 2017–2018 season. He received his MS degree in mechanical engineering from Guelma University and an Executive MBA from the Quantic School of Business and Technology.

John Cook retired from Schlumberger Cambridge in 2018, after 35 years in research and development there. His main focus was with geomechanics and its use in drilling planning and optimization, wellbore instability control, lost circulation, stimulation, sand production, and field and reservoir management. Cook is the author or co-author of many technical papers for SPE and other journals, co-author of a chapter in the SPE-published *Advanced Drilling and Well Technology* handbook, and assignee on a number of patents.

Before joining Schlumberger, Cook worked on materials science and electron microscopy, with a BS degree in natural sciences and a PhD in physics, both from Cambridge University.

Ken Russell has been geomechanics adviser and director of Russell Geomechanics, based in Aberdeen, Scotland, since 2014. He has gained more than 45 years of experience working for a variety of oil service companies in wireline operations, management, and formation evaluation. Since 2000, Russell principally has worked in real-time geomechanics operations and developing acousto-geomechanical applications. He was one of Schlumberger's principal instructors, delivering geomechanics training at operating company locations, training centers, and operational centers worldwide. Through extensive operational and wellsite experience in the North Sea, Europe, Africa, South America, and the Far East, Russell has gained a broad-based knowledge of drilling, production, log data acquisition, analysis, and interpretation that has allowed him to develop and deliver pragmatic solutions to the geomechanical challenges of drilling, sand production, fracturing, and unconventional reservoirs faced by operators. His principal interests include the development and application of acousto-geomechanical techniques for the evaluation of anisotropic formations and fracture systems and the identification and prevention of wellbore instability. Russell holds a BSc (Hons) degree in physics from Aberdeen and has been a member of SPE since 1984.

LOST CIRCULATION: A NEW APPROACH TO AN OLD CHALLENGE

SALIM TAOUTAOU, JOHN COOK, KEN RUSSELL

1. Lost Circulation: Definition, Challenges, and Consequences

Lost circulation has been a challenge in the oil and gas industry since its beginning. The industry has worked tirelessly to solve this challenge, but it still persists and is still one of the most costly factors in drilling and well construction.

Recent advances in geomechanics, drilling techniques, drilling fluids, cementing and lost circulation fluids, logging while drilling, and data analytics, are now assisting the industry to resolve lost circulation challenges more effectively. This book discusses the present understanding of lost circulation and the use of these new techniques.

1.1. Definition and Challenges. Lost circulation occurs when drilling fluid flows into one or more geological formations instead of returning to the surface. It has a global footprint; it is costly and impacts well integrity and productivity. It costs the industry approximately USD 1 billion per year (Al Maskary 2014; Droger 2014). In the south of Mexico, the losses in the Mesozoic Fields have been estimated to cost PEMEX USD 1 million per year (Orellan 2010).

Lost circulation is a serious problem in the oil and gas industry. It is estimated (Economides 1998) that 25% of wells worldwide suffer from lost circulation. In geothermal wells, lost circulation accounts for between 3.5 and 10% of total project costs (Finger 2010).

Lost circulation can occur at different stages during well construction, and can be classified as

- Induced losses, often attributable to poor drilling practices, such as surge and swab, or formation pressure conditions, such as reservoir depletion. Therond (2017) found that 90% of losses are initiated during cementing operations while running the casing/liner and during the pre-job circulation before cement enters the annulus.
- Losses to pre-existing natural fractures. Lost circulation is common in the naturally fractured carbonate formations of the Middle East (i.e., Saudi Arabia, UAE, Oman, Qatar, and Iraq). Losses are observed in different horizons such as the Dammam, Hartha, Shuaiba, Khuff and Umm er Radhuma Formations. In Saudi Arabia, 32% of wells in the Khuff Formation experience wellbore breathing, while 10% experience lost circulation (Ameen 2014).
- Losses to vugs and cavities. Voids in the subsurface created during deposition or as a result of dissolution may act as the recipients of total losses.

The severity of the losses is normally characterized by the loss rate. Seepage losses occur in permeable rocks and in zones with fissures and microfractures; the loss rate is less than 10 bbl/hr. For partial losses, the loss rate can vary from 10 to 200 bbl/hr. Partial losses can be experienced in high permeability rocks and in zones with fissures and fractures. Severe losses occur in highly fractured or vugular formations, where the loss rate can reach 500 bbl/hr. Total losses are simply where there are no returns of the drilling fluid to the surface.

1.2. Consequences. The detrimental effects of lost circulation are felt, not only during well construction, but also on production and economics during the life of the field.

1.2.1. Drilling. If mud is lost to the formation while drilling and the well cannot be kept filled, the hydrostatic pressure exerted by the mud at the wellbore wall is reduced. This favors wellbore instability and the production of breakout cavings. Excess solids in the wellbore can lead to the drilling assembly becoming stuck.

A lost drilling assembly may lead to losing the drilled section or even the whole well. The situation may call for a new casing design; an additional casing string may be added to enable the drilling and complete the well. The downsides of this are to increase the telescopic effect of the well and reduce the size of the completion to be deployed, thus increasing cost, and jeopardizing ultimate production from the well.

If, as a result of drilling fluids being lost into the formation, the pressure in the well is lower than the pore pressure, a well control situation exists, and an influx of fluids can occur. When formation fluids and gases enter the wellbore there is an increased risk of potential discharge to surface, which would create an environmental hazard as well as potential loss of life in the worst-case scenario.

1.2.2. Well Integrity. Mud losses, or lost circulation during cementing operations, may result in the inability to achieve the required downhole hydraulic isolation. Poor cement coverage and failure to achieve the required top of cement compromise the working envelope of the well. Because the casing is not protected by the cement, it will be subject to thermal and mechanical stresses and to corrosion, which can lead to casing failure (i.e., collapse or burst).

When the annulus behind casing is not completely cemented, it will be exposed to downhole fluids migrating from the formation. This can cause trapped pressure in the annuli, commonly characterized as sustained casing pressure (SCP), sustained annulus pressure (SAP), or casing-to-casing annulus pressure (CCA). This is the pressure that occurs in an annulus that rebuilds after having been bled off.

Trapped pressure in the annuli indicates that well integrity is compromised because the primary barrier (cement in this case) is not providing hydraulic isolation. Poor cement coverage and lack of verification will result in costly repairs with a high uncertainty of success.

Cement integrity assurance and corrosion are determined by running acoustic, ultrasonic, electro-magnetic or high-resolution caliper logging tools. Where possible, the primary barrier can be reinstated by perforating the casing and pumping sealant (cement or other materials) to repair the defects.

1.2.3. Production and Reservoir Damage. Drilling fluids lost to the reservoir may cause permeability damage and other skin effects, hindering the production of hydrocarbons.

SCP, CCA, and SAP are of particular concern. They are indications of a failure of one or more barrier elements of the well. The loss of integrity in the well will ultimately lead to an uncontrolled release of fluids, which in turn can lead to unacceptable safety and environmental consequences.

The lack of isolation has a detrimental effect on the well. The casing will be vulnerable to chemical attack from drilling and reservoir fluids. External casing corrosion will occur as a result of aquifer penetration of corrosive brines and carbon dioxide (CO_2). Hydrogen sulfide (H_2S) attack will make the casing brittle and subject to cracks.

In all cases, the well's working envelope imposed by the maximum-allowable surface pressure and the maximum-allowable operating pressure is jeopardized as a result. The well production pressure will be reduced, affecting the overall production of hydrocarbon.

1.2.4. Economics: Unplanned Spending. Lost circulation costs the industry approximately USD 1 billion per year (Al Maskary 2014; Droger 2014). During the drilling phase, these unplanned costs include the cost of the nonproductive time (NPT), the cost of material, and the cost of its associated services. These costs have more impact in high-risk/high-value wells such as high-pressure/high-temperature or deepwater wells, where the cost of NPT is more accentuated as a result of the high rig-operating cost and the logistical challenges.

During cementing operations, unplanned spending includes the cost of lost circulation material treatments and the extra volumes of slurries pumped. Costs also include unplanned logging runs that may be required to verify whether the barrier is achieved and remedial cement jobs that are attempted (mostly unsuccessfully) to fix any defects. Redrilling or unplanned workover operation costs and, finally, the unrealized revenue resulting from lost production may render a project uneconomical.

Summary

- *Lost circulation occurs when drilling fluid, as well as cementing and completion fluids, flow into one or more geological formations instead of returning to the surface.*
- *It has a global footprint affecting 25% of wells worldwide.*
- *Lost circulation costs the industry around $1 billion per year.*
- *It has detrimental economic effects on well and field operations.*
- *Lost circulation causes reservoir damage, therefore hindering the production of hydrocarbon.*

1.3. References.

Al Maskary, S., Halim, A. A., and Al Menhali, S. 2014. Curing Losses While Drilling & Cementing. Paper presented at the Abu Dhabi International Petroleum Exhibition and Conference, Abu Dhabi, UAE, 10–13 November. SPE-171910-MS. https://doi.org/10.2118/171910-MS.

Ameen, M.S. 2014. Fracture and in-situ stress patterns and impact on performance in the Khuff structural prospects, eastern offshore Saudi Arabia. *Marine and Petroleum Geology* **50** (February):166–184. https://doi.org/10.1016/j.marpetgeo.2013.10.004.

Droger, N., Eliseeva, K., Todd, L. et al. 2014. Degradable Fiber Pill for Lost Circulation in Fractured Reservoir Sections. Paper presented at the IADC/SPE Drilling Conference and Exhibition, Fort Worth, Texas, USA, 4–6 March 2014. SPE-168024-MS. https://doi.org/10.2118/168024-MS.

Economides, M. J., Watters, L. T., and Dunn-Norman, S. eds, 1998. *Petroleum Well Construction*, Chap. 5, 135. Hoboken, New Jersey: Wiley-Blackwell.

Finger, J. and Blankenship, D. 2010. *Handbook of Best Practices for Geothermal Drilling*, Sandia National Laboratories, Report SAND2010-6048, Contract No. DE-AC04-94AL85000, US DOE, Albuquerque, New Mexico and Livermore, California (December 2010). https://www.energy.gov/sites/prod/files/2014/02/f7/drillinghandbook.pdf.

Orellan, S., May, R., Bedino, H. et al. 2010. Design of "Anti Surge" Methodology to Mitigate Severe Lost Circulation While Running Non-Conventional Casing / Liner Sizes to Isolate Salt and Clay Domes in Deep Wells in Mexico South. Paper presented at the IADC/SPE Asia Pacific Drilling Technology Conference and Exhibition, Ho Chi Minh City, Vietnam, 1–3 November. SPE-135905-MS. https://doi.org/10.2118/135905-MS.

Therond, E., Taoutaou, S., James, S. G. et al. 2017. Understanding Lost Circulation While Cementing: Field Study and Laboratory Research. Society of Petroleum Engineers. Paper presented at the SPE/IADC Drilling Conference and Exhibition, The Hague, The Netherlands, 14–15 March. SPE-184673-MS. https://doi.org/10.2118/184673-MS.

2. Mechanics of Rocks and Fractures

2.1. Geomechanics Primer. Rocks will deform and eventually fail if the forces acting on the rock exceed the ability of the rock to withstand those forces. Thus, all rock mechanics studies focus on determining the rock mechanical properties of the formations of interest and determining the magnitudes and orientations of the stresses acting on the rock (Jaeger 2007). Of particular interest is determining the properties and stresses in the vicinity of cavities or discontinuities, such as boreholes, perforation tunnels, and natural or hydraulically induced fractures. The accuracy of our knowledge of the rock properties and the stresses acting on the rock will determine the reliability of our predictions of the conditions of wellbore and reservoir pressure that may induce fracturing and lost circulation.

2.1.1. Stresses and Mohr's Circle. In general, there will be one normal and two shear stresses acting on each of the faces of an elemental cube of rock in the subsurface. However, there will always be a unique orientation of that cube such that the shear stresses on the faces of the cube are reduced to zero. Then the full stress tensor describing the in-situ state of stress is given by three orthogonal normal stresses (the principal stresses) and their orientation in space (**Fig. 2.1**).

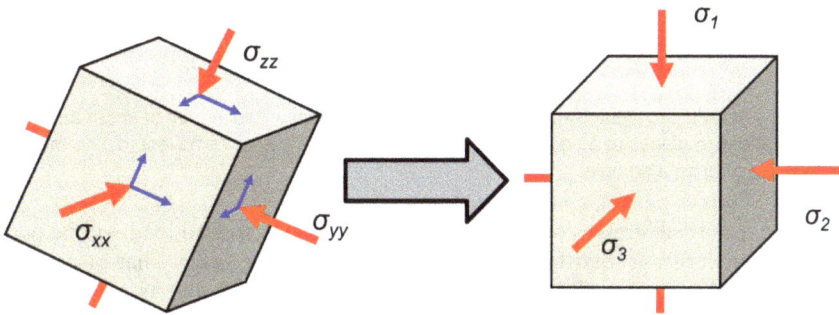

Fig. 2.1—There is always a unique orientation of the generalized state of stress in the subsurface such that the state of stress can be described by three principal normal stresses and their orientation in space. (Normal stresses in red and shear stresses in blue.)

For many situations, one of those principal stresses will be vertical and the three principal stresses will be named the overburden stress and the maximum and minimum horizontal stresses. However, in some areas of the world, for example, near tectonic plate boundaries or around salt domes, the vertical stress may not be the largest stress and the principal stress orientations may be inclined from the vertical. Shear stresses cannot exist in a gas or a fluid (air or water). Therefore, the principal stresses at the earth's surface must be oriented parallel and perpendicular to the earth's surface. Because most stresses in the subsurface are compressional, in geomechanics, compression is considered positive and tension negative (contrary to mechanical engineering convention).

The state of stress in the subsurface is defined by the full stress tensor: the division of the stress state into three principal stresses is only a useful mathematical convenience. Because stress is a force applied over a specific area, the principal stresses cannot be considered as vector forces and resolved by simple trigonometry.

When two unequal normal stresses, σ_1 and σ_3, are applied to a sample of rock, shear stresses are generated within the sample. The locus of all possible combinations of normal (σ_n) and shear stresses (τ) acting on all possible planes that make an angle (θ) between the plane normal to the bigger stress (σ_1), and the plane normal to the smaller stress (σ_3) can be visualized as a circle on a Mohr's circle plot (**Fig. 2.2**).

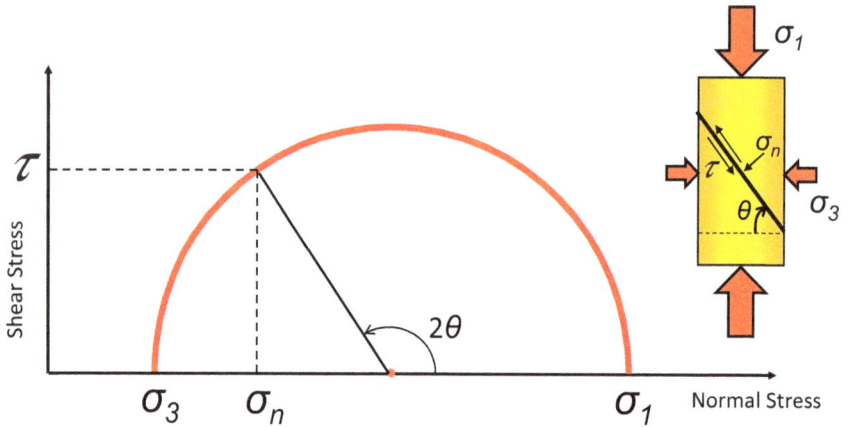

Fig. 2.2—Mohr's circle diagram illustrating the locus of all possible values of shear and normal stress in planes at an angle (θ) between the largest (σ_1) and smallest (σ_3) normal stresses acting on a sample.

When a normal stress is applied to a porous rock, the applied load will be partitioned between the solid frame and the fluid in the pore space. That portion of the load applied to the solid frame is called the effective stress and its magnitude is defined as the total normal stress (σ_n), less a function of the pore pressure (P_p),

$$\sigma_n' = \sigma_n - \alpha P_p. \dotfill (2.1)$$

The factor α is the Biot parameter and, for the elastic deformation of common sedimentary rocks, it takes a value around 0.9. It will be the effective stress and its magnitude in comparison to the strength of the rock that will cause the rock to fail. When considering rock failure, the Biot parameter does not appear in the analysis, and the normal effective stress is given by

$$\sigma_n' = \sigma_n - P_p. \dotfill (2.2)$$

2.1.2. Elastic Rock Properties and Failure Criteria. In the seventeenth century, Robert Hooke discovered the law of elasticity that bears his name and that describes how the extension of an elastic spring varies linearly with the load applied to it. For the elastic deformation of sedimentary rocks, Hooke's law tells us that the stresses and strains are related by elastic moduli. For isotropic materials, only two independent moduli are required. In geomechanics we use the Young's modulus (E) and the Poisson's ratio (v). For other applications, other moduli are used (e.g., shear modulus, bulk modulus, Lamé parameters, or Thomsen parameters). However, these other parameters are related to E and v via simple algebraic relations. Young's modulus is defined as the ratio of the applied stress to the resultant strain in the same axis. Poisson's ratio is the ratio of the radial to the axial strain in one direction normal to the applied stress.

Rock elastic properties can be measured in the laboratory (**Fig. 2.3**) using a triaxial load cell. The sample is placed in a rubber sleeve and submerged in a bath of hydraulic fluid that can exert a confining stress, $\sigma_2 = \sigma_3$. The axial stress (σ_1) is applied by way of steel platens that are part of a stiff loading frame. Axial and radial deformations are measured and reported as millistrains or microstrains, where strain is the ratio of the change in length to the original length. The Young's modulus is simply the slope of the linear part of the axial stress to axial strain data. The Poisson's ratio is the slope of the radial strain to axial strain data. While the applied stresses allow the sample to remain within the linear elastic region, reduction of the axial stress will allow the sample to return to its original dimensions. Whereas, if the axial stress exceeds a certain critical value (the yield strength), other mechanisms, such as internal rearrangement of the grains, microcracking and mineral dissolution, will result in irreversible permanent plastic deformation. Understanding how the rock yields is critical to the understanding of fracture formation and fracture propagation. If the axial stress is further increased, the rock will eventually rupture and fail. The peak stress at failure is often taken as the compressive strength of the rock at that confinement.

For a granular material such as rock to fail in compression, the magnitude of the shear stress, ($|\tau|$) on the plane of failure must not only exceed the cohesion (S_0) holding the grains together, but also must exceed the friction generated by the effective normal stress (σ_n) acting perpendicular to the plane. The Mohr-Coulomb failure criterion is written as

$$|\tau| = S_0 + \left(\mu.\sigma_n'\right), \quad \dots\dots\dots\dots\dots\dots\dots\dots\dots\dots\dots\dots \text{(2.3)}$$

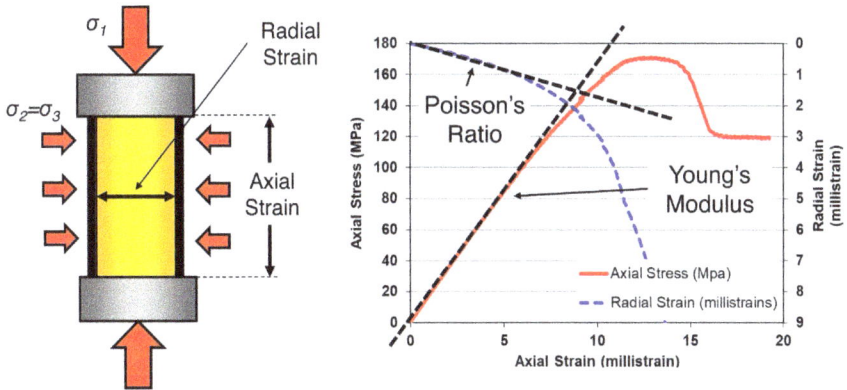

Fig. 2.3—Measurements of the axial and lateral strains resulting from loading a sample in a triaxial load frame. The applied stresses and measured strains are used to derive the elastic parameters, Young's modulus and Poisson's ratio.

where μ is the coefficient of internal friction in the rock. To obtain the failure parameters [unconfined compressive stress (UCS) and angle of internal friction, ϕ] in the laboratory, a number of triaxial loading experiments at different confining pressures are performed. At each confinement the rock is loaded to failure and the confining stress and peak stress at failure are plotted graphically as circles on a Mohr's diagram (**Fig. 2.4**). The line tangential to all the circles represents the failure envelope.

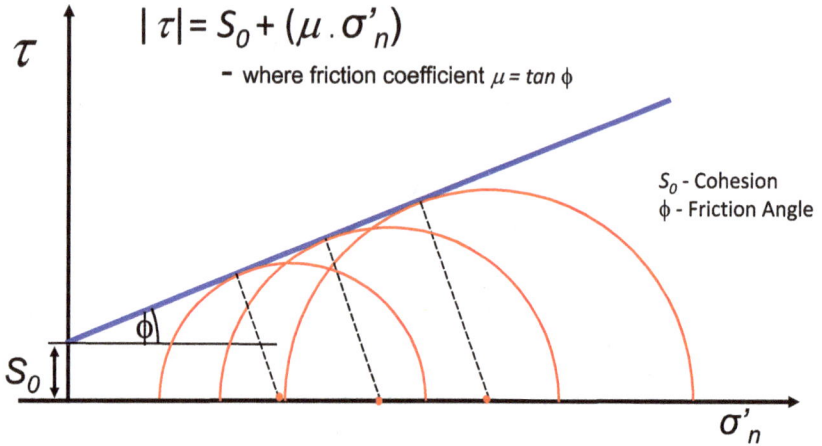

Fig. 2.4—Mohr's circle diagram at three values of confining stress. The failure envelope (blue line) delivers the values of cohesion (S_0) and the angle of internal friction (ϕ).

The results may also be plotted on an axial stress (σ'_1) vs. confining stress (σ'_3) plot (**Fig. 2.5**). In this case, the Mohr-Coulomb failure criterion may be rewritten as

$$\sigma'_1 - N_\Phi \cdot \sigma'_3 = C_0, \dots\dots\dots\dots\dots\dots\dots\dots\dots\dots\dots (2.4)$$

where the slope of the line is the triaxial stress factor, $N_\phi = (1 + \sin \phi)/(1 - \sin \phi)$, and the intercept on the y-axis at zero confinement (C_0) is the UCS.

2.1.3. State of Stress and Pore Pressure in the Subsurface. Once the mechanical properties of the rock have been established, the geomechanics engineer needs to evaluate the state of stress where that rock is situated in the subsurface. The overburden stress (σ_v) at a depth (h) in the subsurface is simply the sum of the bulk density (ρ_b) of all the material above that depth times the acceleration due to gravity (g),

$$\sigma_v = g \int_{-h}^{0} \rho_b dz, \dots\dots\dots\dots\dots\dots\dots\dots\dots\dots\dots (2.5)$$

where the density of water is around 0.433 psi/ft and the density of overburden rock is around 1 psi/ft. Remember that in modern field developments, the overburden can be a complex function of water depth, well deviation, and formation changes (**Fig. 2.6**).

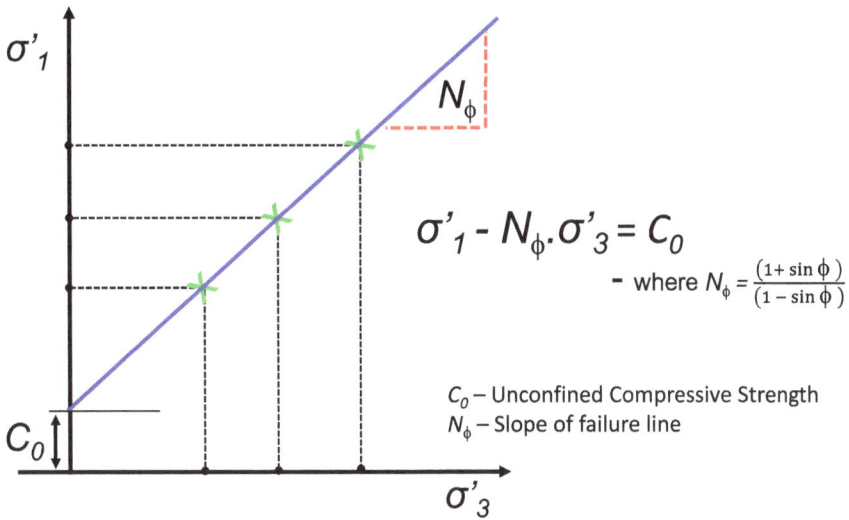

Fig. 2.5—The same data as Fig. 2.4 is redrawn as a σ_1' vs. σ_3' plot. The intercept on the y-axis delivers the value of UCS.

$$\sigma'_1 - N_\phi \cdot \sigma'_3 = C_0$$

$$\text{- where } N_\phi = \frac{(1 + \sin \phi)}{(1 - \sin \phi)}$$

C_0 – Unconfined Compressive Strength
N_ϕ – Slope of failure line

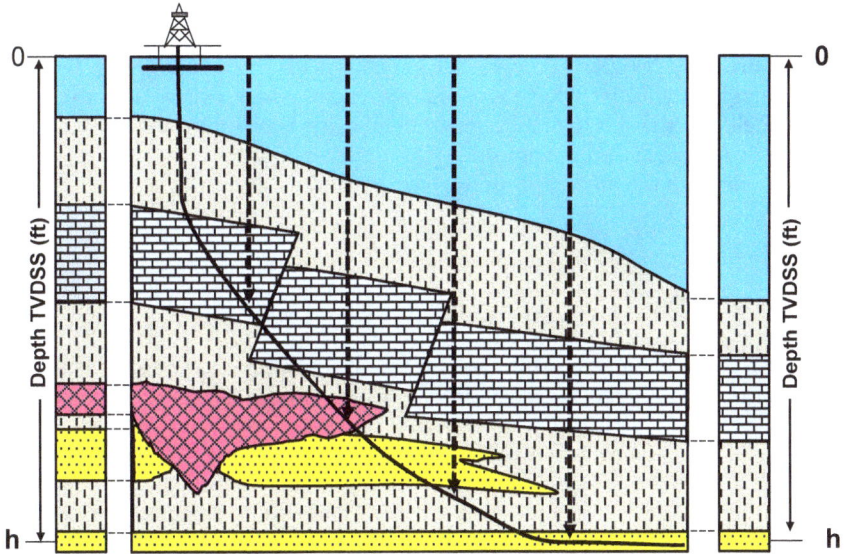

Fig. 2.6—Diagram illustrating the need to take account of the changing overburden formations and depths of water in deviated wellbores when calculating overburden stress.

The overburden stress is not necessarily the largest principal stress in the subsurface. According to the Andersonian scheme, the stress regime in the subsurface is denominated as normal faulting, thrust faulting, or strike-slip (wrench) faulting, depending on whether the magnitude of the vertical stress is respectively the

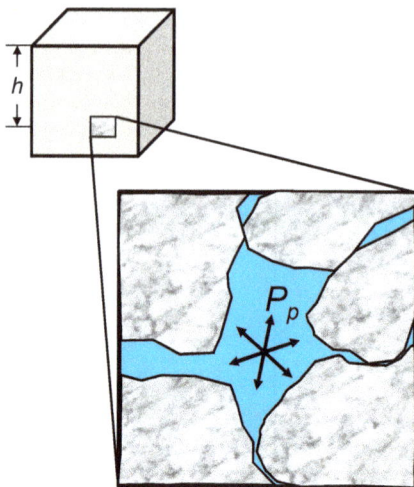

Fig. 2.7—The pressure in pore fluids acts equally in all directions and is single valued.

largest, the smallest, or the intermediate stress of the three principal stresses in the subsurface.

The pore space between grains in rock contains fluids or gases (**Fig. 2.7**). Because fluids and gases effectively have zero shear strength, pore pressure takes the same value in all directions. If hydraulic connectivity exists between the depth of interest and the surface, then the pore pressure (P_p) is determined as the product of the depth (h), the density of the fluid in the pore space (ρ_f), and the acceleration of gravity (g):

$$P_p = \rho_f gh. \dotfill (2.6)$$

When water fills the pore space, this pore pressure is called the hydrostatic pressure. For common formation waters, a pressure gradient of 0.433 psi/ft is often used. However, abnormal pore pressures are observed worldwide and various techniques to measure and estimate abnormal pore pressures have been developed (Huffman 2003).

In porous and permeable formations, pore fluids can be drawn from the formation into formation tester devices and the formation pressures measured with a pressure gauge. Analysis of the measured pressure data can deliver formation pressure and permeability. Optional arrangements of packers and pumps can deliver azimuthal permeability, microfracture stress measurements, and fluid typing, among others.

In porous but impermeable formations, such as shales, extracting fluid from the formation to measure pressure is not a viable option. Various techniques to transform trends of measured properties (e.g., velocity, density, resistivity, and drilling rate) to pore pressure are available (Huffman 2003). These techniques have been developed to account for the effects of shale undercompaction, as well as fluid-expansion effects (i.e., heating, hydrocarbon generation, and shale and carbonate diagenesis), lateral transfer and tectonic loading (Sayers 2010).

Estimates of pore pressure are critical to the drilling engineer to design drilling programs for appropriate mud weights and casing setting depths, for example. However, we must remember that the pore pressures derived from empirical transforms are only estimates, not measurements of pressure, and appropriate procedures need to be in place to mitigate against the uncertainties in these estimations.

The principal source of horizontal stresses in the subsurface is by way of gravitational loading (**Fig. 2.8**). An elemental volume of rock in the subsurface will be compressed by the overburden and, through the Poisson's ratio of the rock, will want to expand horizontally. However, the deformation of rock in the subsurface is constrained laterally by the adjacent rock; therefore, horizontal stress is generated.

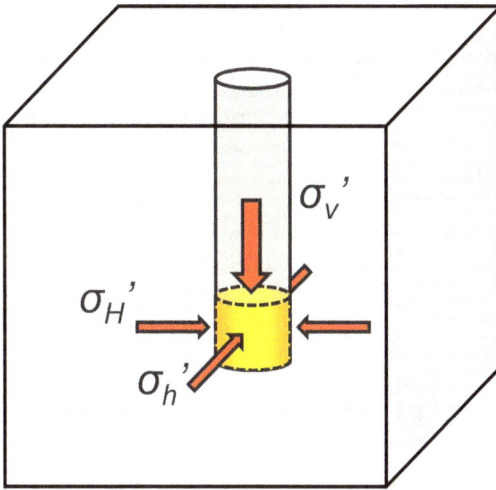

Fig. 2.8—Diagram illustrating how horizontal stresses are generated through gravitational loading.

For isotropic rocks, the portion of the magnitude of the horizontal stresses resulting from gravitational loading can be estimated using a simple uniaxial strain model:

$$\sigma'_h = \sigma'_H = \frac{v}{(1-v)}\sigma'_v. \quad \dots\dots\dots\dots\dots\dots\dots\dots\dots\dots\dots \quad (2.7)$$

In addition to the horizontal stress generated by gravitational loading, terms caused by the strains generated by tectonic plate movement and the effects of increasing (or decreasing) temperature in the subsurface should be added. The general form of the horizontal stress equations then becomes

$$\sigma'_h = \frac{v}{1-v}\sigma'_v + \frac{E}{1-v^2}\varepsilon_h + \frac{vE}{1-v^2}\varepsilon_H + \frac{E}{1-v}\alpha_T \Delta T, \quad \dots\dots\dots\dots\dots \quad (2.8)$$

$$\sigma'_H = \frac{v}{1-v}\sigma'_v + \frac{E}{1-v^2}\varepsilon_H + \frac{vE}{1-v^2}\varepsilon_h + \frac{E}{1-v}\alpha_T \Delta T, \quad \dots\dots\dots\dots\dots \quad (2.9)$$

where E is Young's modulus; ε_h and ε_H are the horizontal tectonic strains in the minimum and maximum stress directions, respectively ($\varepsilon_H > \varepsilon_h$); and α_T is the linear elastic coefficient of thermal expansion of the rock. ΔT is the increase in temperature of the rock resulting from burial, proximity to geothermal bodies, or to the injection of hot (or cold) fluids in the rock.

The horizontal stresses in the subsurface are rarely equal. The inequality can be accounted for by imbalanced tectonic strains caused by folding, faulting, or the proximity to other structural features such as salt diapirs. Many rock types (most notably shales) exhibit vertical-to-horizontal anisotropy of the elastic rock properties. Similar treatments for the evaluation of the horizontal stress state exist for anisotropic rocks (Jaeger 2007).

The formulation used to estimate the magnitudes of the horizontal stresses should then be validated by comparison to measured values from leakoff, microfracture or minifracture test data. The orientation of the horizontal stresses can be estimated by evaluation of subsurface structure on seismic data. However, the present-day orientation of horizontal stress is preferably verified by earthquake data or observation of wellbore failure artifacts observed on caliper, image, or other wellbore log data (Heidbach et al. 2018).

2.1.4. Driller's Fracture Gradient and the Mud Weight Window. When selecting a mud weight to use, drillers will select a value such that the well pressure is greater than pore pressure to avoid kicks (fluid influx) and is less than the driller's fracture gradient to avoid fluid losses. These two limits are often referred to as the safe mud weight window and are often presented as a "ppfg" plot in the drilling program (drillers also allow a safety margin to these values, referred to as kick tolerance). Measuring and estimating pore pressure to determine the lower limit of mud weight has already been discussed. To determine the upper limit to mud weight, drillers perform leakoff tests at the start of each drilled well section. The leakoff value obtained is assumed to be the formation fracture gradient at the shoe. This value is used to calibrate empirical relations, such as Matthews and Kelly or Eaton, to predict the fracture gradient for the rest of the section to be drilled (**Fig. 2.9**).

This technique provides a useful method for drillers to estimate the likely upper limit of well pressure and select a mud weight to avoid mud losses or lost circulation over the whole of the subsequent well section. In contrast to this approach, the geomechanics engineer wants to use his knowledge of the stresses and rock properties to predict which interval in the subsequent well section may cause lost circulation. Because it is the minimum principal stress that holds natural fractures closed, and the full stress tensor that determines the initiation of drilling-induced fractures at the wellbore wall, geomechanics engineers will use their prediction of the minimum principal stress as the geomechanics fracture gradient.

To obtain the minimum principal stress at the casing shoe, it is now common practice for drillers to perform extended leakoff tests (Hauser 2020). This test involves pumping sufficient volume at surface to breakdown the formation, creating an induced fracture and propagating the fracture beyond the near wellbore stresses. When pumping is stopped, the resulting analysis of the pressure data will determine at what pressure the fracture closes. That fracture closure pressure represents the best measurement of the minimum principal stress (fracture gradient) in the subsurface. Drillers are sometimes reluctant to use the extended leakoff test method because it

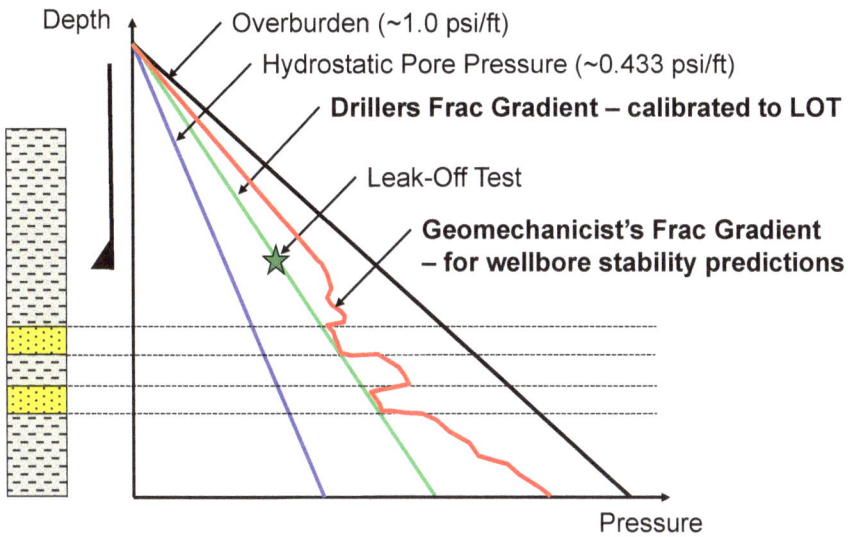

Fig. 2.9—Graphical representation of how drillers use leakoff test data to guide their choice of fracture gradient (green). Geomechanics engineers use the minimum principal stress (red) of the full stress tensor as their fracture gradient.

involves opening a fracture at the casing shoe. However, how can you measure a fracture closure pressure, if you have not already opened a fracture? Most modern bottomhole drilling assemblies include a real-time annular pressure measurement. The digital pressure data acquired from downhole provides higher resolution and accuracy and allows for better pressure-decline analysis and determination of the closure pressure/minimum principal stress. Using such drilling assemblies allows for drilling the casing shoe, performing the leakoff test, and drilling ahead without the requirement to pull out of hole.

In addition to the safe mud weight window used by drillers, geomechanics engineers will use their estimates of rock properties and subsurface stresses to calculate the mud weight limits to ensure a stable wellbore. The shear failure gradient is the lower limit of mud weight to avoid shear failure at the wellbore wall and subsequent production of cavings. The tensile failure gradient is the upper limit of mud weight to avoid tensile failure and the initiation of drilling-induced fractures. The combined mud weight window defines the safe and stable envelope of mud weight to drill a specified well trajectory.

In deepwater or high-pressure wells the safe and stable mud weight window is likely to be very narrow, creating wellbore instability, mud loss, and lost circulation challenges. In such circumstances, reliable geomechanical modeling is critical.

2.2. Geological Aspects of Natural Fractures. Nearly all rocks contain natural fractures (Fossen 2012). The global movement of tectonic plates, erosion, sedimentation, and other mechanisms such as earthquakes and salt diapirism ensure that over geologic time, rock is in a constant state of deformation and failure. Natural fractures divide and separate the rocks in the subsurface into matrix blocks. The flow of fluid within the subsurface is therefore governed by the relative magnitudes of the porosities and permeabilities of the rock matrix and of the natural fractures that divide it.

Dyke (1995) comments that in a well "production levels of thousands of barrels a day can flow from very short naturally fractured intervals." Equally, high permeability naturally fractured intervals can cause "massive losses of drilling mud flowing from the wellbore into the surrounding formations". An understanding of the complexity of how natural fractures are formed, their connectivity, and whether they are open, closed, or partially cemented is critical to managing lost circulation challenges during well construction.

2.2.1. Characterization of Natural Fractures. The term *fracture* is used to describe a wide variety of features observed in the subsurface, from microfractures in individual grains to large-scale fracturing and faulting at the mountain building scale. Individual fracture widths may vary from the micron to the centimeter scale and higher.

As sedimentary layers in the subsurface become folded in a thrust belt or, as they become draped over a rising salt diaper, fissures and fractures are created (Dashti 2009; Cosgrove 2015; Stearns 1964). Flexing of the layers causes changes in the local state of stress. For example, high horizontal compression at the bottom of the layer in **Fig. 2.10a** causes the local state of stress to be thrust faulting, creating horizontal opening fractures. Whereas, at the top of the layer, the local state of stress is normal faulting, creating vertical opening fractures.

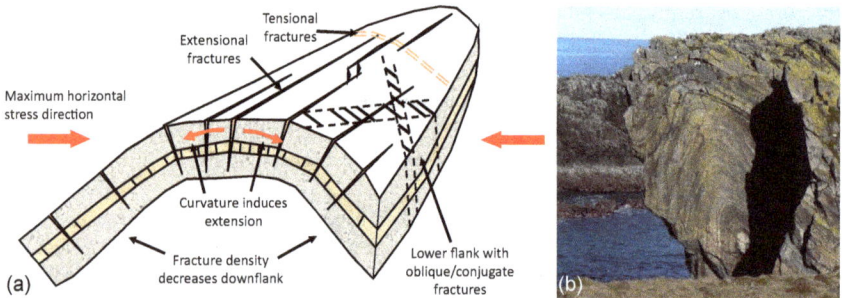

Fig. 2.10—(a) Diagram illustrating the stress changes and changes in failure mode within a sedimentary layer subjected to folding, after Jadoon (2005); (b) folded and fractured quartzose schists, Kinnaird Head, northeast Scotland; photo courtesy of Ken Russell.

Opening or dilational fractures are created when the stress normal to the plane of the fracture becomes tensile. In sedimentary formations the bedding planes separating layers create narrow permeable fractures (**Fig. 2.11**). The effects of heating and diagenetic shrinkage in coal seams create sets of face and butt cleats that act as conduits for coalbed methane production. Open fractures (**Fig. 2.12**) may be created in carbonates because of the dolomitization of limestones. In volcanic situations, wide veins or dykes are created by injections of magmatic material.

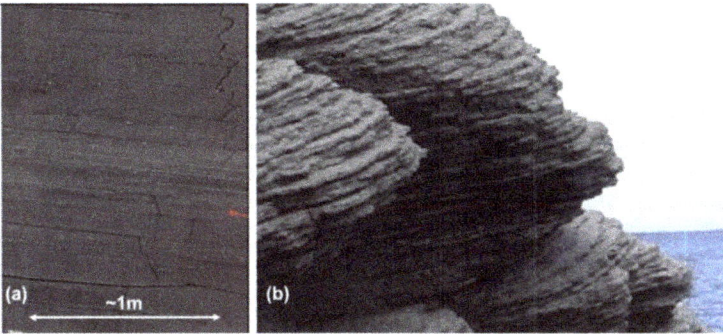

Fig. 2.11—(a) Horizontal bedding plane fractures in the Bazenhov oil source rock in Tyumen, Russia. The bedding planes are crossed by a vertically propagating fracture that is exploiting the weaknesses in the bedding planes, after Platunov et al. (2013); and (b) layered and naturally fractured shale in outcrop, after Mohaghegh (2013).

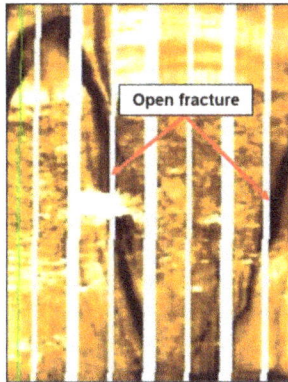

Fig. 2.12—Large-aperture open fracture and associated fracture damage observed on electrical borehole image log from southwest Iran; after Dashti et al. (2009). Note that the apparent aperture seen on image logs is a function of sensor resolution and near wellbore stresses and may not represent the in-situ fracture aperture.

Shear fractures are created when the shear stress in the plane of the fracture is sufficient to overcome the cohesion and friction in the rock. The failure causes displacement of the two faces of the fracture parallel to the fracture plane. Faults are the result of large shear displacements. Individual shear fractures tend to have narrow apertures and low permeability. However, the combination of multiple intersecting shear fractures within a fault plane may be sufficient to create a permeable fracture corridor (**Fig. 2.13**).

Sorkhabi (2014) provides a comprehensive natural fracture characterization scheme that can be summarized as follows:

- Type of fracture (dilational or shear), whether it is open, partially or fully filled, and properties of the filling material

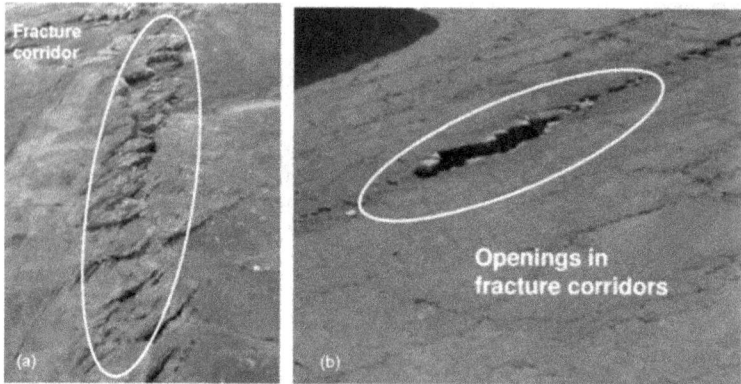

Fig. 2.13—(a) Multiple en-echelon shear fractures, and (b) high permeability flow channel in fracture corridors, Clare Basin, Ireland; after Zhang and Koutsabeloulis (2010).

- Association of fractures with particular lithology, structure, geologic age, and in-situ state of stress
- Number of fracture sets and their relative ages and modes of creation
- Individual fracture length, height, strike direction, and dip
- Fracture set spacing, intensity, and connectivity
- Porosity and permeability, related to equivalent fracture aperture and asperity (roughness)
- Fracture stiffness, describing how fracture deforms under changing conditions of fluid pressure and stress

2.2.2. Mineralization of Natural Fractures. The void spaces created by natural fractures are likely to fill with formations fluids. Depending on the chemical composition of those fluids and the conditions of temperature, pressure, and fluid flow experienced by the fracture over geologic time, some amount of precipitation of dissolved minerals (e.g., quartz, calcite, or anhydrite) on the faces of the fracture is expected to occur. The degree of mineralization will have a significant effect on the porosity and permeability of the fracture (Olson et al. 2009). Multiple layers of mineral precipitation may occur as a variety of fluid types flow through the fracture at different geological ages. As precipitation of minerals proceeds, bridging across the fracture may occur (**Fig. 2.14a**), and eventually the fracture may become completely filled (**Fig. 2.14b**). Mildly acidic formation fluids may have the opposite effect: leaching and etching the fracture faces, increasing fracture apertures, and creating vugs and other dissolution features.

When a natural fracture or dyke is created by the injection of magmatic material, the liquid magma will fill the void space. As it cools, the liquid magma tends to crystalize, sealing the fracture entirely. Fracture-filling minerals may be stronger or weaker than the matrix rock. **Fig. 2.15** shows an example of how the weaker sandstone matrix rock has been weathered and eroded leaving the harder fracture-filling minerals standing proud of the outcrop.

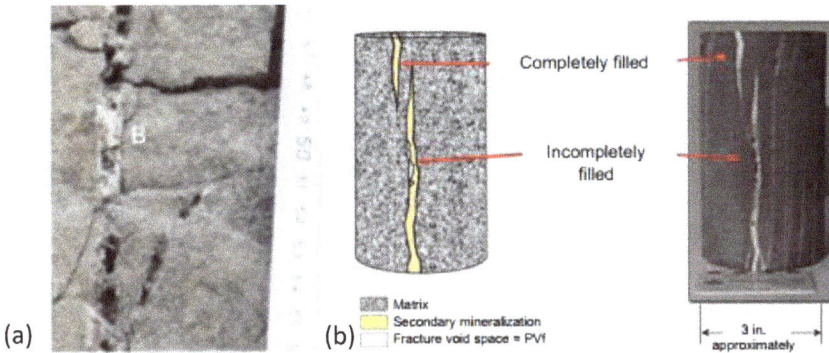

(a)   (b)

Fig. 2.14—(a) Natural fracture in Triassic La Boca Sandstone, Mexico, illustrating quartz cement bridging and filling at B, scale in mm, after Laubach et al. (2010); and (b) core and schematic from the Lower Devonian Forillon Limestone, Quebec, Canada, showing natural fractures filled and partially filled with calcareous minerals; from Piedrahita and Aguilera (2017).

Fig. 2.15—Weathered and fractured sandstone where the fracture-filling material is harder than the sandstone matrix, Stokes Bay, Kangaroo Island, Australia; photo courtesy of Ken Russell.

2.3. Rock Mechanical Properties of Natural Fractures. Natural fractures are initiated at stress concentrations around microfractures or defects in the rock structure. Once a fracture is initiated, the stress concentration can result in three different types of displacement (**Fig. 2.16**): Mode I (opening perpendicular to plane of fracture—tensile failure), Mode II (sliding in plane of fracture parallel to fracture propagation—shear failure), Mode III (tearing in plane of fracture perpendicular to fracture propagation—shear failure). The Mode I opening is of most interest for lost circulation applications. However, the other modes become of interest when considering fracture re-orientation and tortuosity in the near wellbore or when an induced fracture encounters a natural fracture system.

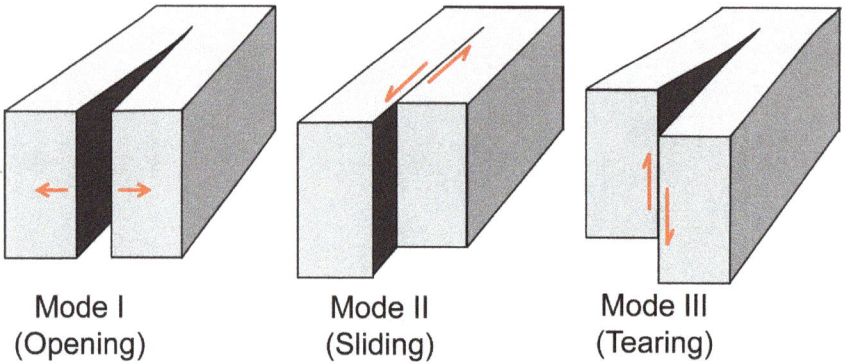

Mode I   Mode II   Mode III
(Opening)   (Sliding)   (Tearing)

Fig. 2.16—Classification of natural fractures by mode of displacement.

2.3.1. Fracture Toughness and Fracture Propagation. A measure of the stress singularity at the fracture tip is the stress-intensity factor, K. For a simple Mode I slit geometry,

$$K_I = -\sigma\sqrt{\pi X_f}, \dots\dots\dots\dots\dots\dots\dots\dots\dots\dots\dots\dots\dots (2.10)$$

where σ is the stress acting perpendicular to the fracture faces and X_f is the fracture half-length. As a fracture is opened, stress will be concentrated at the tip of the fracture. The longer the fracture, the higher the stress intensity at the tip; once initiated, a fracture is easier to propagate.

The fracture will propagate when the stress-intensity factor (K_I) exceeds a critical value known as the critical-stress-intensity factor (K_{IC}), or fracture toughness. When the fracture is propagating, net pressure in the fracture is given by:

$$\left(P_f - \sigma_3\right) = \frac{K_{IC}}{\sqrt{\pi X_f}}, \dots\dots\dots\dots\dots\dots\dots\dots\dots\dots\dots\dots\dots (2.11)$$

where σ_3 is the minimum principal stress acting perpendicular to the fracture faces and P_f is the pressure in the fracture (**Fig. 2.17**). For perfectly elastic material, fracture toughness is a material property with values varying from 1 MPam$^{1/2}$ for limestones

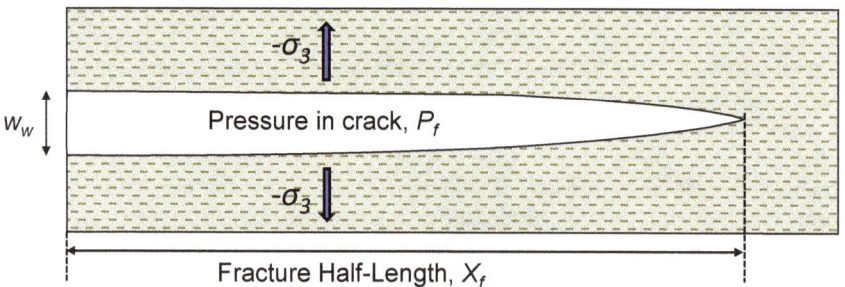

Fig. 2.17—Schematic of idealized fracture illustrating Mode I opening.

up to 4 MPam$^{1/2}$ for granite (Siren 2012). Various experimental techniques have been devised to measure the value of fracture toughness in the laboratory (Funatsu 2015).

2.3.2. Fracture Width. Drilling fluids lost to fractures in the subsurface contain solids. Not only solids added to the mud system to control mud rheology, but also a mixture of crushed cuttings and cavings. The diameter of these particles in relation to the fracture aperture will determine whether these particles are admitted and transported into the fracture or whether they bridge and plug the fracture aperture. Under the simplifying assumptions of Mode I geometry, a propagating fracture would have a fracture width,

$$w_w = \left(P_f - \sigma_3\right) \times \frac{4\left(1 - v^2\right)X_f}{E}, \dotfill (2.12)$$

at the wellbore wall (Fig. 2.17). Note that fracture width is inversely proportional to the Young's modulus—the stiffer the rock, the narrower the fracture.

2.3.3. Fracture Compliance. The fracture planes in a naturally fractured formation tend to be aligned with each other. Any increase in effective stress normal to the plane of open or partially filled fractures will cause the fracture width to reduce (i.e., the fracture set is more compliant in the orientation normal to the fracture plane than the orientation parallel to the fracture plane). Thus, the acoustic velocity normal to the fracture plane will be slower than the velocity parallel to the fracture plane.

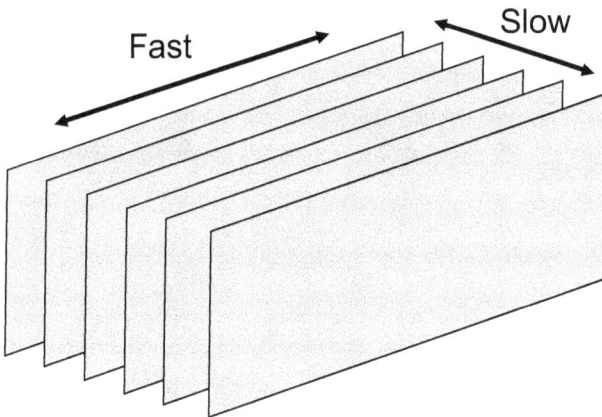

Fig. 2.18—Naturally fractured systems are more compliant in the orientation normal to the fracture planes, resulting in slower acoustic velocities.

Many seismic and wellbore acoustic wave analysis techniques have been developed to identify the excess compliance caused by natural fracture systems and determine whether the fractures are likely to be open or closed (Sayers 2010). Fracture orientation is determined by computing the azimuth of the fast and slow acoustic wave propagation.

An open fracture contains fluids or gases. Because fluids or gases cannot support shear stress, the local orientation of the principal stresses at the fracture must be normal and parallel to the fracture faces. For Mode I opening fractures, the minimum principal stress will be oriented normal to the fracture face. For small fractures of minimal width and limited areal extent, the effect of stress re-orientation may be insignificant. However, for large natural fracture networks and faults, the re-orientation of stresses to be normal and parallel to the fault plane can have significant effect on fracture permeability, mud losses, and wellbore stability for wells drilled through or in the vicinity of the fault.

2.3.4. Natural Fracture Networks. Sedimentation and burial of rock at elevated temperature and stress, will cause permanent deformation and compaction. Subsequent reduction in stress and temperature caused by erosion and uplift may open weak bedding planes and create extensive natural fracture networks. **Fig. 2.19** shows a predominantly orthogonal joint fracture set in low-porosity Devonian-age sandstones. Thin layers of volcanic dust have created weak bedding planes that are easy to split; the resulting flagstones have been used in construction for many centuries.

Fig. 2.19—Well-developed joint sets on flagstones at St. Mary's Chapel, Caithness, Scotland; © Mike Norton* (2008). The predominantly orthogonal joint sets are crosscut by a diagonal fracture set (Sorkhabi 2014).

2.4. Induced Fractures. Since the introduction of hydraulic fracture stimulation in the 1940s and 1950s, a large body of literature has been published concerning the initiation, propagation, and fluid flow within induced fractures (Economides and Nolte 2000). Of particular interest to lost circulation challenges is the understanding of how drilling-induced fractures are initiated in intact rock and how those fractures interact with natural fractures in the formation.

2.4.1. Drilling-Induced Fractures at the Wellbore Wall. In the laboratory we know the forces acting on the rock sample, and it is relatively easy to predict failure.

*Permission is granted to copy, distribute and/or modify this document under the terms of the GNU Free Documentation License, Version 1.2 or any later version published by the Free Software Foundation; with no Invariant Sections, no Front-Cover Texts, and no Back-Cover Texts. A copy of the license is included in the section entitled *GNU Free Documentation License.*

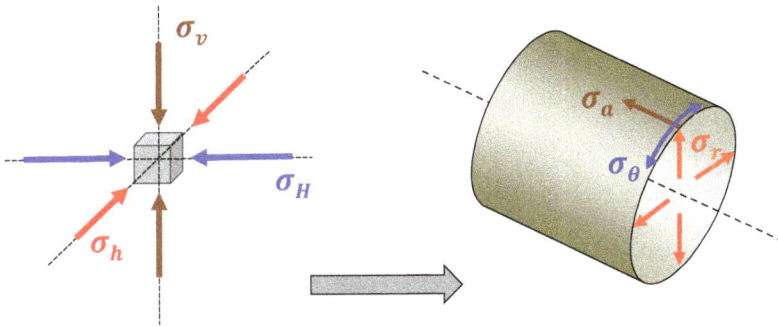

Fig. 2.20—Diagram illustrating how the principal in-situ stresses in the far field must re-orient to be parallel and perpendicular to the wellbore wall.

However, before we can predict failure in a wellbore, we need first to estimate the stresses in the formation at the wellbore wall. Replacing rock with drilling fluid alters the stress state in the near wellbore. Because fluids cannot support shear stresses, the three principal stresses at the wellbore wall must act parallel and perpendicular to the surface of the wellbore (**Fig. 2.20**). The orientation of the wellbore with respect to the far-field stresses will determine how the stress magnitudes and orientations are altered at the wellbore wall.

For the simple cylindrical geometries employed in the oilfield, the alteration of the stress state due to the presence of wellbores or perforation tunnels can be described by Kirsch's equations (Jaeger 2007). For the special case of a vertical wellbore, the axial, radial, and tangential (or hoop) effective stresses at the wellbore wall can be approximated as follows:

Maximum tangential effective stress, $\sigma'_{\theta max} = 3 \times \sigma'_H - \sigma'_h - \left(P_p - P_w\right),$ . . . (2.13)

Minimum tangential effective stress, $\sigma'_{\theta min} = 3 \times \sigma'_h - \sigma'_H - \left(P_p - P_w\right),$ . . . (2.14)

Radial effective stress, $\sigma'_r = P_p - P_w,$ . . . . (2.15)

Axial effective stress, $\sigma'_a = \sigma_v - P_p.$ . . . . (2.16)

The tangential stress at the wellbore wall varies from a maximum value oriented parallel to the minimum horizontal stress, to a minimum value oriented parallel to the maximum horizontal stress. Notice that the magnitudes of the tangential and the radial stresses both contain the pressure in the wellbore (P_w) as a parameter. If well pressure changes cause the radial effective stress to increase by an amount, the tangential effective stress decreases by the same amount. Therefore, any increase or decrease of well pressure resulting from mud weight changes, circulation, swab, or surge, directly affects the stresses acting on the rock at the wellbore wall. Management of wellbore pressure by the driller becomes the dominant factor in managing and maintaining wellbore stability.

Failure Mode Transition Pressures

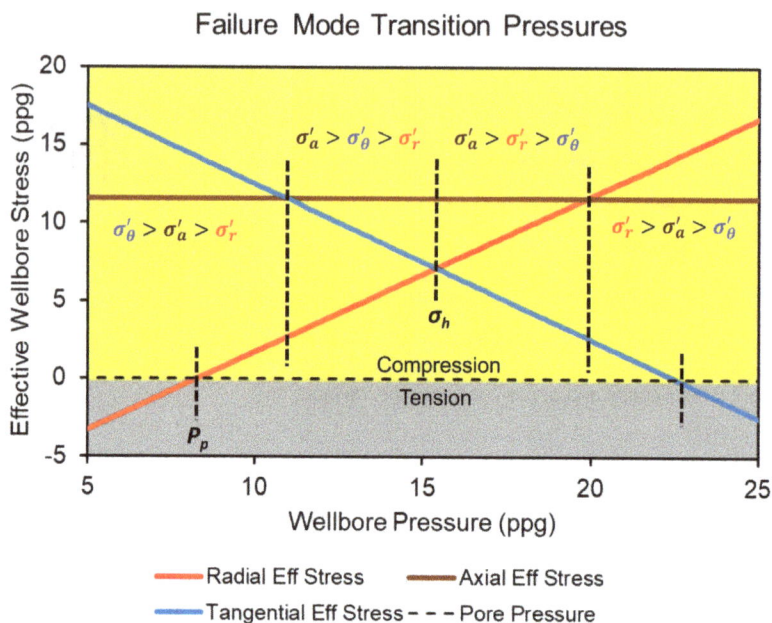

Fig. 2.21—Diagram illustrating how wellbore pressure affects the magnitudes of the three principal stresses at the wellbore wall for a simplified far-field stress state. Failure mode transition pressures are marked as vertical dashed lines. See text for detailed explanation and Bratton (2020) for the methodology used.

If we also assume that the two horizontal stresses are the same magnitude (although generally they are not the same), we can illustrate how, for a given far-field stress state, the magnitudes of the three principal stresses at the wellbore wall vary with well pressure (**Fig. 2.21**).

The plot (Fig. 2.21) naturally divides into two areas. One area (shaded yellow) is where wellbore effective stresses are compressive, and the other area (shaded grey) is where wellbore effective stresses are tensile. For low wellbore pressures (less than the pore pressure), the radial effective stress is negative/tensile. If that tensile stress is greater than the tensile strength of the rock, the rock will fail in tension. Because the plane of the tensile failure must be perpendicular to the radial tensile stress, the failure will act circumferentially around the wellbore and thin sharp cavings will spall into the wellbore. Equally, for high wellbore pressures where the tangential effective stress is negative/tensile and greater than the tensile strength of the rock, the rock will fail in tension and vertical drilling-induced fractures on either side of the wellbore will result. The plane of this induced fracture must be perpendicular to the tangential tensile stress so the induced fracture will propagate radially away from the wellbore.

Warpinski et al. (1982) found that although lithology contrasts had an effect on fracture height growth, the principal factors controlling height growth were stress

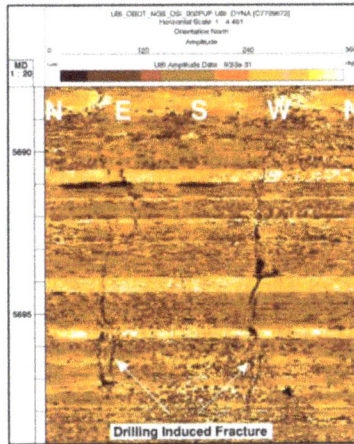

Fig. 2.22—Ultrasonic borehole image of drilling-induced fractures from the Tullich Field, North Sea (depth scale in feet). The trace of the fractures on the image demonstrates the complexity of the failure and the influence of formation lithology on the position of the fractures on the wellbore wall (Russell et al. 2006).

contrasts and weak interfaces (**Fig. 2.22**). He found that only a 300–400-psi contrast in stress between layers was sufficient to arrest fracture height growth. Further work by Fisher and Warpinski (2012) provides the mineback experimental data and the analytical equations that control height growth.

If the wellbore pressure required to initiate the vertical drilling-induced fracture is greater than the minimum horizontal stress, that fracture will continue to propagate away from the wellbore. If that induced fracture intersects with an open natural fracture system or other high permeability system, severe mud losses may occur. However, if a drilling-induced fracture is created with a wellbore pressure less than the minimum horizontal stress, the induced fracture cannot continue to grow beyond the near wellbore and wellbore 'breathing' may be observed when pumps are shut down and wellbore pressures fall at connections.

It has been common practice in the industry to assume that the only mechanism for creating drilling-induced fractures is when the pressure in the wellbore is sufficiently high for the tangential stress to become tensile and greater than the tensile strength. However, there are other compressional mechanisms that can cause the rock to fail and create permeable cracks, fissures, and fractures in the near wellbore that could act as conduits for mud losses. If we refer to Fig. 2.21 again, we see that in the yellow-shaded area there are four separate compressional stress regimes. Each regime is defined by the order of the magnitudes of the effective stresses at the wellbore wall. We know from Mohr's circle analysis that shear failure occurs in the plane containing the biggest and the smallest principal stresses. We also know that the plane of failure is oriented at an angle greater than 45 degrees from the plane perpendicular to the biggest stress to the plane perpendicular to the smallest stress. Conventionally, we think of shear failure and the production of breakout cavings occurring at low mud weights. This occurs when the tangential stress is the largest and the radial stress is the smallest stress at the wellbore wall. The shear fractures created by this failure mode/stress regime tend to form near parallel to the wellbore wall, so do not create permeable paths for mud losses. Whereas the fractures that may form in the other three possible compressional stress regimes radiate away from the borehole, creating conductive pathways for mud losses to the formation or any natural fracture network.

When we see wellbore image log data acquired after a hydraulic fracture operation, we often observe very complex failure patterns on the wellbore wall; a mixture of tensile- and shear-induced fracturing (Bratton 2020). Our mental image of a symmetrical and perfectly planar bi-wing fracture propagating away from the wellbore is destroyed by the evidence. Real fractures—induced or natural— have complex geometries.

There is a timing element to failure. When a rock is subject to applied stress, the rock will fail in its weakest geometry. Thus, if the shear stress at the wellbore wall exceeds the shear strength of the rock prior to the tensile stress exceeding the tensile strength, then the rock will fail in shear (creating a narrow aperture fissure/fracture in the process). Conversely, if the tensile stress exceeds the tensile strength of the rock before the shear stress exceeding the shear strength, then the rock will fail in tension (creating a wider aperture fracture). Which failure occurs first depends on the relative magnitudes and orientations of the far-field stresses with respect to the wellbore orientation. Multiple wellbore failure modes are possible and observed on image log data. Not all drilling-induced fractures will be wide aperture, high-permeability features, and difficult to plug with lost circulation material. Some may be tortuous, narrow-aperture shear fractures that plug easily with mudcake forming solids already in the mud system. Nevertheless, once an induced fracture at the wellbore wall is propagated beyond the influence of the near-wellbore stress regime, it is likely that the induced fracture will be failing in tension in Mode I (Bratton 2020). For a normally stressed regime, Olson et al. 2009 found that opening Mode I failure is likely to precede shear failure in the far field for conditions in which the unconfined UCS of the formation rock is greater that the effective vertical stress.

However, once failure starts (the production of cavings or the creation of drilling-induced fracturing) the simplifying assumptions of cylindrical geometry no longer apply, and the predictions of the Kirsch solution for stresses at the wellbore wall are no longer valid. Because of the adverse environmental conditions in wellbores, it is not possible to acquire all the measurements that would be needed to precisely model the stresses and predict the multiple failure modes that would occur in wells that have already failed.

2.4.2. Induced Fractures due to Reservoir Pressure Depletion. It has long been understood that as reservoir pressure reduces during production, the magnitudes of the horizontal stresses also reduce (**Fig 2.23**). Because the surface of the earth is free to move, it is often assumed that as fluid is withdrawn from the reservoir, the overburden simply subsides and the magnitude of the overburden stress at reservoir depth can be assumed to remain constant. However, because rock in the subsurface is constrained laterally, the horizontal stresses in the reservoir reduce as pore pressure reduces.

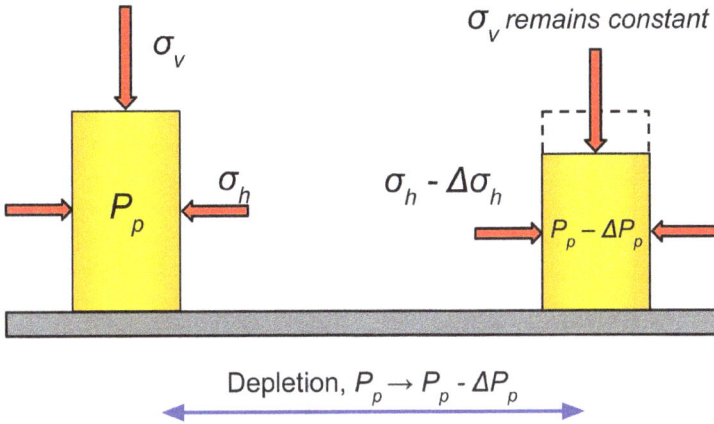

Fig. 2.23—Schematic illustration of how horizontal stresses reduce when formation pore pressure is reduced.

Using Hooke's law for elastic behavior, and assuming that the overburden stress does not change ($\Delta\sigma_v = 0$), and that there is no horizontal strain ($\varepsilon_h = 0$), then the changes in total and effective horizontal stresses can be calculated as

$$\text{Change in total horizontal stress, } \Delta\sigma_h = \alpha \frac{1-2v}{1-v} \cdot \Delta P_p, \quad \dots\dots\dots\dots\dots \text{(2.17)}$$

$$\text{Change in effective horizontal stress, } \Delta\sigma_h' = -\alpha \frac{v}{1-v} \cdot \Delta P_p. \quad \dots\dots\dots\dots \text{(2.18)}$$

Note that while the reservoir pressure depletes by an amount, ΔP_p, the total horizontal stress decreases, but the effective horizontal stress increases. So, it should come as no surprise to us that the rocks in and around depleted reservoirs fracture as reservoir pressures decrease. The horizontal reservoir stress path is defined as

$$\gamma_h = \frac{\Delta\sigma_h}{\Delta P_p}. \quad \dots\dots\dots\dots\dots\dots\dots\dots\dots\dots\dots\dots\dots\dots\dots\dots\dots\dots\dots\dots\dots\dots\dots \text{(2.19)}$$

Values of γ_h can be measured experimentally on core samples in the laboratory. However, the effect of reservoir and overburden formation structure may cause arching and the stress path at the reservoir scale may be different from the laboratory value (**Fig. 2.24**).

The effect of subsurface structure and faulting may mean that the stress paths in the two principal horizontal stress directions may be different. As depletion proceeds, the differential reduction in stress may cause the orientation of the horizontal stresses to rotate or simply flip 90 degrees. The change in stress orientation will affect the orientation of any induced fracturing. Modern reservoirs are being produced from ever smaller and more complex geometries. Full 3D modeling of the effects of structure on pore pressure and stress evolution is recommended.

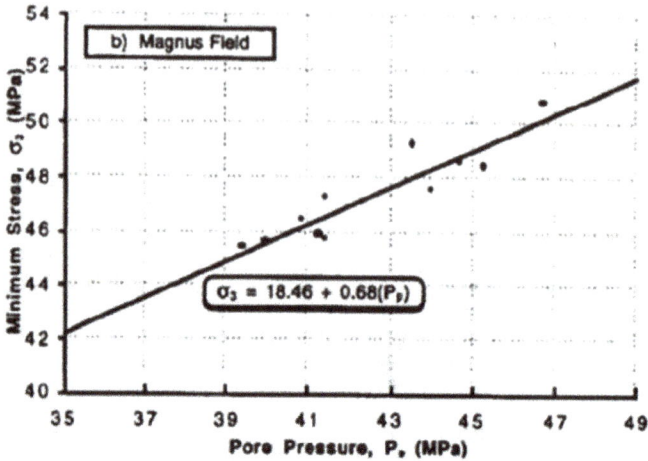

Fig. 2.24—Reduction in total horizontal stress as pore pressure depletes in the Magnus Field, North Sea (Addis 1997).

Before a reservoir starts to deplete, the subsurface stresses are in equilibrium. To maintain equilibrium as hydrocarbons are extracted and reservoir pressure depletes, the horizontal stresses lost in the reservoir interval must redistribute and add themselves to the horizontal stresses immediately above and below the reservoir (**Fig. 2.25**). This effect requires 3D modeling and is well known in the mining industry (Esterhuizen 2008) where much effort is expended in pillar design for headings and roadways to mitigate against roof failure.

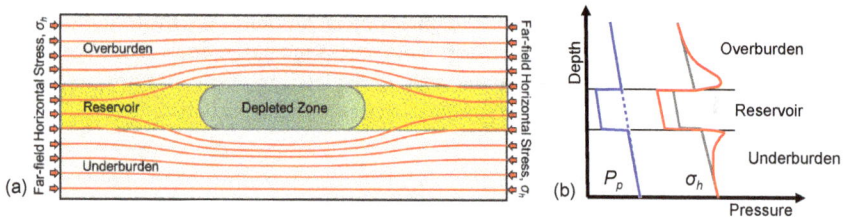

Fig. 2.25—(a) Schematic illustrating how horizontal stresses must redistribute to above and below the reservoir as pore pressure depletes, (b) graphical illustration of how the original horizontal stress magnitude (grey) changes in the reservoir and in the immediate overburden and underburden (red).

Simple Mohr's circle analyses illustrate how shear failure can induce fracturing within and immediately above and below the depleted reservoir rocks. In the reservoir (**Fig. 2.26**), as pore pressure reduces, the effective horizontal stresses increase, but not by the same amount as the pressure depletion. Thus, the Mohr's circle increases in diameter and moves to the right. For sufficient pressure depletion, the circle will eventually touch the failure line and shear fractures will be produced in the reservoir. These fractures may improve permeability but may affect well integrity

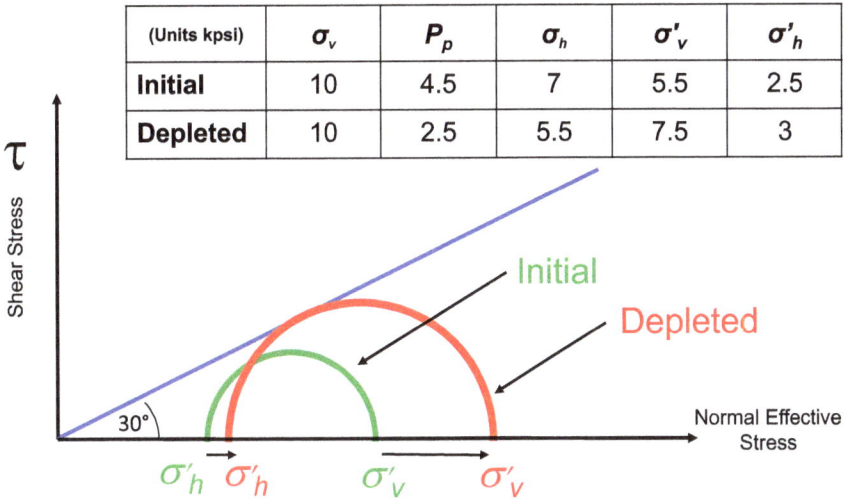

(Units kpsi)	σ_v	P_p	σ_h	σ'_v	σ'_h
Initial	10	4.5	7	5.5	2.5
Depleted	10	2.5	5.5	7.5	3

Fig. 2.26—Example of Mohr's circle analysis in the depleting reservoir interval. As pore pressure depletes the horizontal and overburden, effective stresses increase in magnitude, moving the circle to the right. The diameter of the circle also increases until such time as the circle touches the failure envelope and failure is initiated.

and alter the orientation of maximum permeability, affecting fluid flow in water injection projects.

The sealing formation above the reservoir is often an impermeable shale. Although pressure is depleted in the reservoir, the pressure in the sealing shale does not deplete, implying that the effective overburden stress in the shale does not change (**Fig. 2.27**). However, the effective horizontal stress in the shale only continues to increase as the stress lost in the depleting reservoir redistributes. If the original state of stress was normal, then as the reservoir depletes, the Mohr's circle diameter in the shale initially reduces as the effective horizontal stress approaches the effective vertical stress. However, increasing reservoir depletion will cause the effective horizontal stress in the shale to continue to increase and surpass the effective vertical stress, creating a thrust-fault regime. Eventually, the circle may grow sufficiently that it meets the failure line. At that point, a network of new, clean, low-angle shear fractures will be created in the shale that covers the whole of the top surface of the reservoir.

This situation is critical to the driller intending to drill into a depleted reservoir. As the drillbit approaches to within a few feet of the top of the reservoir, it may encounter this fracture network, and severe to total losses may occur. From the driller's point of view, it may appear that mud is being lost to the low-pressure permeable reservoir rock, but the losses may actually be to a fracture network in the sealing shale. In this case, the appropriate loss-mitigation procedure would be to pump fracture-blocking materials such as fibers, gels, or sized-particles, rather than pumping filtercake-forming additives to plug pore throats in permeable reservoir rock. See Chapter 5 for a discussion of lost circulation materials and fluids and their uses.

While drilling an injector well in to a heavily depleted field in the North Sea, many incidents of severe mud losses were encountered (Russell et al. 2005). Multiple

(Units kpsi)	σ_v	P_p	σ_h	σ'_v	σ'_h
Initial	10	4.5	9	5.5	4.5
Depleted	10	4.5	14	5.5	9.5

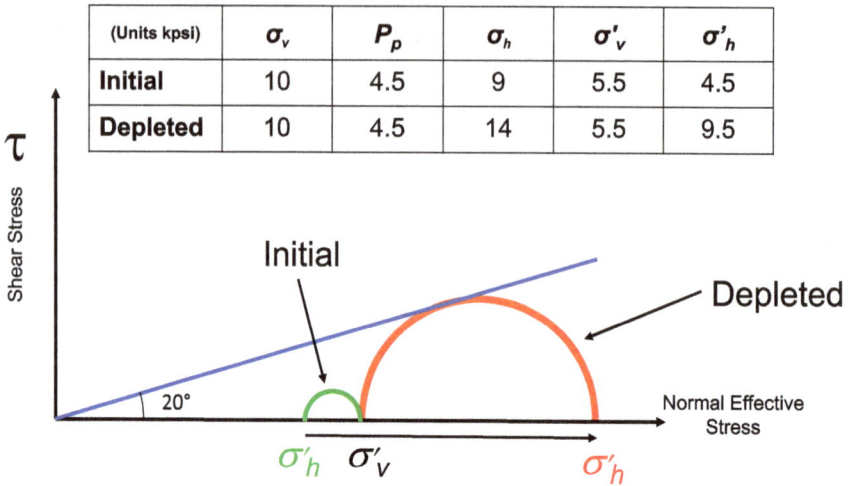

Fig. 2.27—Example of Mohr's circle analysis in the immediate overburden shale. Because the pore pressure in the shale does not change, the effective overburden stress does not change with depletion in the reservoir. The effective horizontal stress only increases until such time as the circle touches the failure line and fractures in the shale immediately above the reservoir are created.

trips were made to condition hole and cure losses. It was noticed that while running in-hole, reaming to alleviate tight-hole conditions initiated losses when the bit crossed the boundary from the sealing shale to the depleted reservoir. When pulling out of hole, overpulls requiring reaming initiated losses when the top stabilizer crossed the same boundary but in the other direction. It soon became obvious that the losses were not to permeable reservoir, but that crushed cavings and cuttings were bridging the fracture network in the sealing shale, creating reduced hole diameter. Each time the bit or top stabilizer crossed the boundary, the seal was being broken and losses to the fracture network were re-initiated.

2.5. Fluid Flow and Particle Transport in Fractures. The analysis of fluid flow from a wellbore in a mud loss or lost circulation situation reduces to a material balance equation. The volume of mud lost from the wellbore is the product of the loss rate multiplied by the length of time the losses were experienced. Some of that volume of mud is lost to the formation (fluid loss through the faces of the fracture), the balance opens the rock to create volume (length, width, and height) in the induced fracture or in opening and propagating the fractures in a connected natural fracture system.

Naturally fractured formations are often considered dual-porosity systems. Fluid flow in the fractures is modeled as a 2D planar system with flow parallel to the plane of the fractures, whereas fluid flow within the matrix is modeled conventionally as a 3D system (Nelson 2001). Fluid flow between the two systems (leakoff) is then modeled depending on the difference in pore pressures, rate of fluid flow, and the relative permeabilities of the two systems.

2.5.1. Mud Rheology. It is well known that the rheological properties of the mud system greatly affect the volume and rate of mud loss during lost circulation events. Modern muds are designed to have low shear strength and high viscosity when the mud is stationary. Whereas, when circulating at high shear rates, the mud should have low viscosity. After the shear stress overcomes an initial yield stress value, the ratio of shear stress to shear rate obeys a power-law relationship. This is called Herschel-Bulkley or yield-power-law behavior. While drilling, these characteristics of the mud are continuously monitored.

Majidi et al. (2010) developed a mathematical model to describe fluid losses of yield-power-law fluids in natural fractures. Their model makes a number of simplifying assumptions, including a nondeformable fracture of constant aperture and infinite length without fracture plugging or fluid leakoff to the formation. They found that as the fluid-invasion front expands within a natural fracture, the shear stresses and shear rates in the mud decrease. When the shear stress in the fluid reaches the yield stress, further expansion of the fluid front is halted. Thus, the yield stress of the fluid controls the maximum amount of fluid that can be lost to the fracture. For a given fracture type and geometry, muds with higher yield stress values lose significantly less volumes than muds with lower-yield stress values. They also found that shear thinning or reduced viscosity of the mud may increase the rate of mud losses but did not affect the ultimate volume lost.

2.5.2. Mud Losses to the Formation—Leakoff. Mud loss to the formation will depend on the areal extent of any newly induced fracture and the areal extent of any preexisting natural fractures opened by the wellbore pressure. The rate at which mud is lost to the formation depends on three mechanisms: the rate at which filter cake is built on the fracture face, the relative permeability of the mud filtrate to the formation, and the compressibility of the formation fluids. All three of these rates reduce in proportion to the inverse of the square root of time. Ideally, these three mud-loss mechanisms should be measured on field samples, especially in the presence of mineralization within the natural fracture system.

The evaluation of leakoff parameters is relatively simple to conduct under static conditions in the laboratory. Under the dynamic conditions of a mud loss or lost circulation event, the effects of filter-cake erosion and fluid degradation at rapidly changing shear rates and temperatures is challenging to evaluate (Economides and Nolte 2000). The difference between the pressure in the fracture and the pressure in the formation is a controlling factor. Frictional losses, flow rates, and flow-regime changes will influence the profile of pressure along the length of the fracture.

Fracturing experiments carried out on large-scale shale samples using a true-triaxial load frame (Suarez-Rivera et al. 2013b), illustrate how the fracturing fluid is lost, not only through the faces of the newly created fracture, but also through the bedding planes opened during the fracturing process (**Fig. 2.28**).

2.5.3. Hydraulic Conductivity. The hydraulic conductivity of natural fractures with rough surfaces and minimal mineralization or cementation can be very sensitive to the effective stress acting normal to the fracture plane (**Fig. 2.29**). While the pressure in the fracture (P_f) remains much lower than the minimum principal stress (σ_h), the natural fracture will remain closed. However, the asperity on the faces of the fracture will result in some initial residual hydraulic conductivity, and fluid loss rates

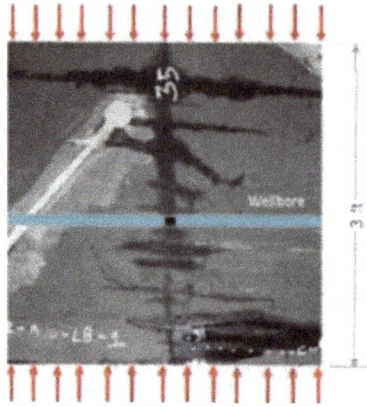

Fig. 2.28—View of large-scale (3 ft x 3 ft x 3 ft) hydraulically fractured shale sample (Suarez-Rivera et al. 2013b). The sample is loaded vertically, and a vertical fracture is created from the horizontal (blue) wellbore. Invasion of fracture fluid is seen along the vertically induced fracture and along the invaded bedding planes.

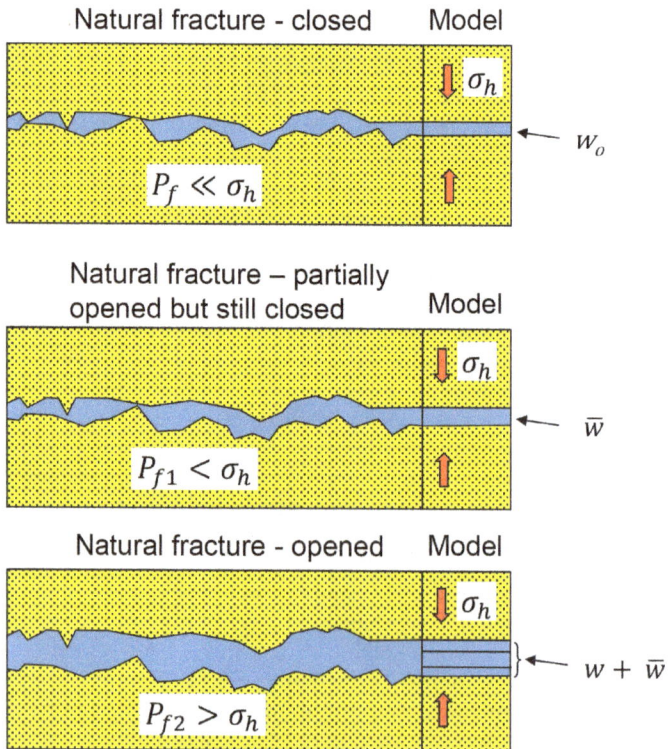

Fig. 2.29—Diagrammatic representation of changing fracture aperture as the pressure in the fracture increases; after Kresse and Weng (2013). The equivalent fracture aperture for use in fracture permeability modeling is demonstrated on the right.

into the natural fracture system will be low; the initial hydraulic aperture of the fracture (w_o) is small. As the fluid pressure in the fracture (P_{f1}) increases and approaches σ_h, the hydraulic aperture (w) will increase, increasing hydraulic conductivity. While the faces of the fracture continue to carry some of the stresses normal to the fracture plane, leakoff rates will increase but remain relatively low. Once fluid pressure in the fracture (P_{f2}) exceeds σ_h, the faces of the fracture will be displaced, and the effective hydraulic conductivity will be the sum of the hydraulic aperture ($\bar{w}$) and the mechanical opening (w) resulting in much-higher fracture permeability and leakoff rates. The resulting change in fluid-loss rate from the wellbore may be dramatic.

2.5.4. Particle Transport. Much theoretical and experimental work has been conducted to understand the admittance of proppant to narrow-slot hydraulic fractures (Economides and Nolte 2000). For drilling fluids with low solids content, an average fracture width of around two times the diameter of the biggest particles is sufficient to admit particles to the fracture. For higher-solids-content fluids, an average fracture width of three times the particle diameter is required. For fracture apertures smaller than this, the larger particles will tend to bridge and plug the fracture aperture. Subsequent drilling fluid flowing into the fracture will pile up behind the bridge and dehydrate and plug the fracture.

Leakoff through the faces of the fracture will build mudcake as mud filtrate is lost to permeable matrix rock. As the fluid flows further along the fracture and more mud filtrate is lost to the matrix rock, the solids content will increase, increasing the likelihood of bridging and plugging.

2.5.5. Interaction of Drilling-Induced and Natural Fracture Systems. When drilling-induced fractures encounter closed natural fracture systems, increased fluid loss and lost circulation may occur. Kresse and Weng (2013) present modeling to simulate complex induced-fracture network propagation in a formation with preexisting closed natural fractures. Critical to modeling and mitigating against lost circulation, is whether the drilling-induced fracture simply crosses the natural fractures it encounters or whether the drilling-induced fracture is arrested, and the natural fracture system opens, diverting fluid and creating high fluid-loss rates (**Fig. 2.30**).

Modeling of the branching of a drilling-induced fracture into a natural fracture system must account for the complex fluid flow, pressure evolution, fracture deformation and propagation, and leakoff rates into both the natural fracture and the matrix. Branching is a dynamic process that must couple stresses, fluid pressures, fracture-tip propagation, and leakoff rates into the opened, partially closed and closed intervals of the invaded natural fractures.

Experimental work by Suarez-Rivera et al. (2013a) using large-scale true triaxial load frames was aimed at understanding the degree of network complexity created during hydraulic stimulation of shales (**Fig. 2.31**). They found that in the presence of high horizontal stress contrasts, prolific fracture branching between weak bedding planes dramatically improved permeability and led to improved reservoir drainage.

Fig. 2.32 illustrates how a rising salt diapir has folded and fractured the overlying carbonate formation. At least two major periods of fracturing can be identified; a northwest/southeast fracture set overlays the major axial fracture set trending west-northwest/east-southeast. By analyzing the fracture densities, fracture lengths,

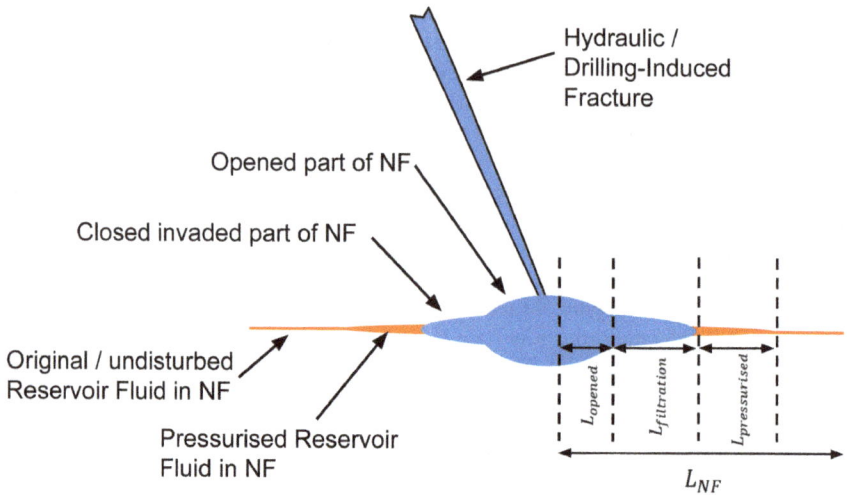

Fig. 2.30—Diagram illustrating the geometry of the intersection of a drilling-induced fracture and a natural fracture (NF). Modeling of the leakoff rates and growth rates of each of the zones in the NF are required to model whether the induced fracture will be arrested or not; after Kresse and Weng (2013).

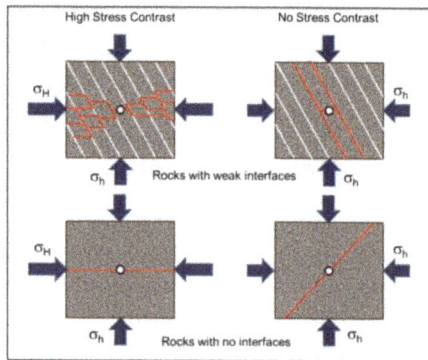

Fig. 2.31—Induced fracture geometry as a function of the presence and orientation of weak bedding planes and the orientation and contrast between in-situ stresses. High-stress contrast results in increased branching; from Suarez-Rivera et al. (2013a).

and relative connectivity of the fracture set interactions (**Fig. 2.33**), it is possible to map the preferred direction of fluid flow through such a fracture complex (Laubach et al. 2019).

2.6. Vugs and Solution Features. During the formation of carbonate rocks, small cavities may be formed. These cavities are called vugs and are often lined with crystals. The secondary porosity created by vugs may or may not be connected and therefore may or may not contribute to the overall permeability of the rock.

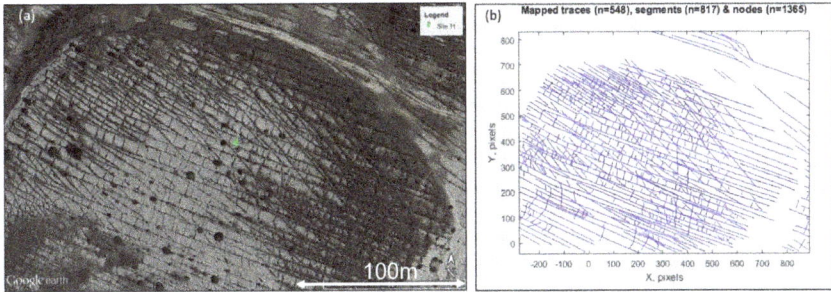

Fig. 2.32—(a) Aerial view of fractured Upper Cretaceous carbonates overlying the Mena salt diaper in the Basque-Cantabrian Basin, northern Spain; from GoogleEarthPro, (b) digital trace of fractures; from Kingsley (2018).

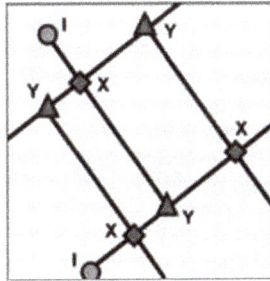

Fig. 2.33—Diagram illustrating the types of fracture interaction; splays (Y), intersections (X), and isolations (I); after Sanderson and Nixon (2015).

Vuggy fractures form when the formation waters flowing through a fracture or joint in a low-permeability rock react chemically to disssolve the matix rock. Once the dissolution process starts, the increased permeability attracts more fluid flow, enhancing the aperture of the vug. In outcrop, these features are sometimes referred to as 'dinosaurs' because of their appearance with a long neck and long tail (**Fig. 2.34**). The excess porosity created is often referred to as secondary porosity and because of its quasispherical shape is relatively incompressible during depletion.

The mode of creation of vugular features implies that they form well-connected and effectively infinite permeability conduits. When a wellbore encounters these features, any well pressure greater than formation pore pressure will result in severe or total mud losses.

Li et al. 2004 present a method to estimate the permeability of fractured basement rocks using wireline image logs and sonic data. They divided the fracture network it to two classes. The primary network was made up of long, hydrothermal solution-enhanced fractures with fracture widths of a few millimeters up to more than a meter; they provided significant porosity and permeability. The secondary network secondary network was made up of short, narrow-aperture discrete fractures with widths less than 0.1 mm. These fractures were more common but more likely to be subject to closure because of stress changes (**Fig. 2.35**).

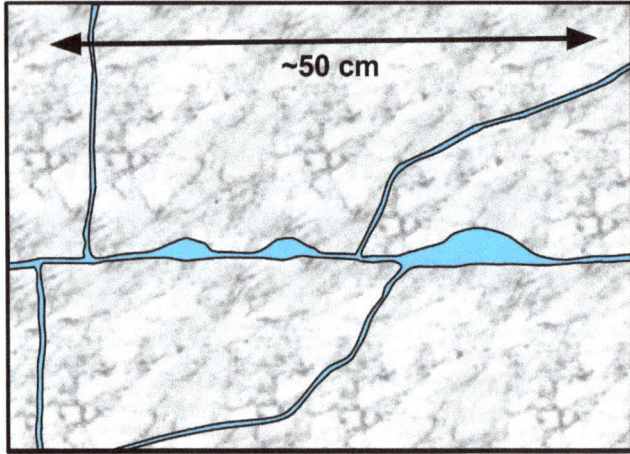

Fig. 2.34—Sketch of solution-enhanced vugs created by flow through joints in limestone.

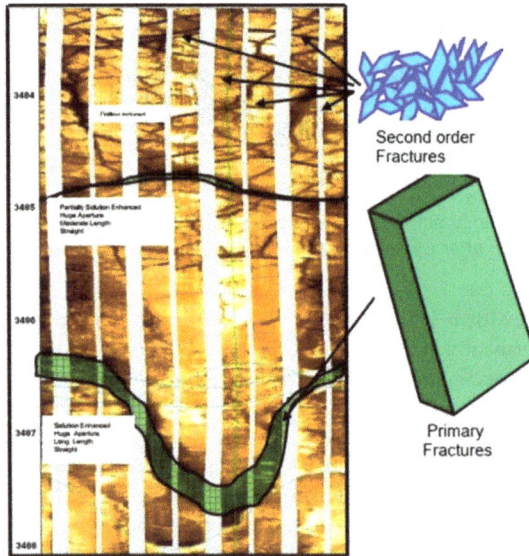

Fig. 2.35—Electrical borehole image log showing the primary network of solution-enhanced fractures and the secondary network of narrow aperture fractures (after Li et al. 2004).

Fracture aperture (F_a) and fracture density (F_d) corrected for well trajectory were estimated from the electrical image logs. Fracture permeability (K_f) was estimated using a simple relationship:

$$K_f = a \times F_d \times F_a^3 \quad \dots\dots\dots\dots\dots\dots\dots\dots\dots\dots\dots\dots\dots\dots\dots (2.20)$$

The factor *a* was used to calibrate the fracture permeability to mud loss rates and, subsequently, to production-flow rates.

Fig. 2.36 shows an example of a high-permeability interval in a very productive well. The blue dots in Track 1 represent the estimation of K_f from the image log data. Track 2 shows the estimation of permeability from the acoustic Stoneley data, and Track 3 shows the strong chevron reflections on the waveform data indicating an open fracture network.

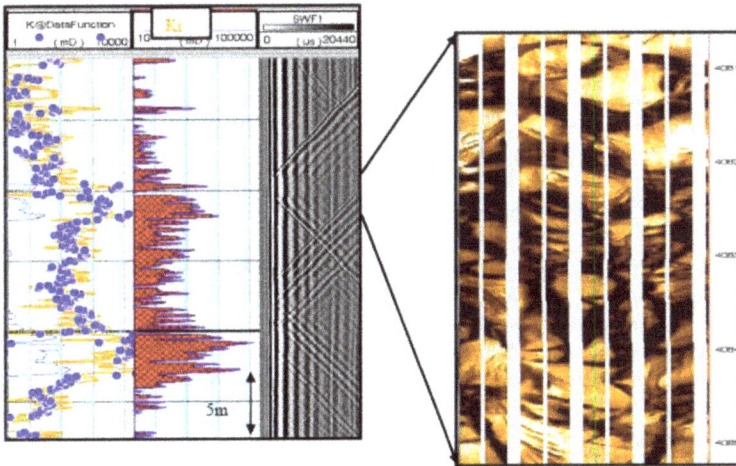

Fig. 2.36—Example illustrating the estimation of fracture permeability from electrical borehole image logs and wireline sonic data (Li et al. 2004).

Carbon dioxide in the atmosphere mixes with rainwater to create mild carbonic acid. **Fig. 2.37** shows how mild carbonic acid waters can create solution-enhanced

Fig. 2.37—Carbon dioxide solution-enhanced features in a limestone near Dubrovnik, Croatia (Shutterstock 2021; Wirestock Creators/Shutterstock.com).

features in carbonate rocks. These features are commonly at the centimeter scale. However, subsurface caverns at scales from 10 m to 100 m may also be encountered while drilling.

Summary

In recent years, the focus on hydraulic fracture stimulation has greatly enhanced our understanding of the geological and geomechanical properties of natural fractures. Nearly all rocks in the subsurface are naturally fractured to some extent. The ease of inducing mud losses and lost circulation will depend on the fluid flow properties (e.g., permeability, aperture, degree of cementation, and compliance) of the fracture. In addition, knowledge of a formation's rock properties and the subsurface stresses acting upon it is necessary to understand the behavior of fractures and the mechanisms causing mud losses or lost circulation. That knowledge will act as a guide in designing the most appropriate mitigation procedures.

- *Fractures can be divided into opening (Mode I) fractures or shear fractures (Mode II and Mode III). Opening (Mode I) fractures can form extensive large-aperture fracture systems and act as the primary recipient of mud losses. Shear fractures (Mode II and Mode III) do not open but may form low-permeability, narrow-aperture systems.*
- *It is the stress concentration at the fracture tip that allows the rock to fail and the fracture to propagate. Far from near-wellbore-stress concentrations, fractures open perpendicular to the minimum principal stress. Once the pressure in the fracture exceeds the minimum principal stress, the fracture faces separate mechanically, fluid can flow unrestricted into the fracture, and fracture propagation is enhanced. The reactivation of preexisting natural fractures depends on its resistance to frictional forces, the pressure in the fracture and its orientation with respect to the subsurface stresses.*
- *The driller's choice of mud weight and management of wellbore pressure will determine the magnitudes of the near wellbore stresses. If well pressure sufficiently alters the stress state, then drilling-induced fracturing is initiated. The drilling-induced fracture may create a permeable conduit for mud losses to the formation or to a connected fracture system.*
- *The reduction in reservoir pressure during production will change the state of stress in both the reservoir and the immediate overburden and underburden formations. For sufficiently large changes of stress, fracturing may be initiated in the reservoir and in the bounding formations. The fracture network created in the caprock immediately above the reservoir may be sufficiently permeable and well connected so that total losses occur as the drillbit approaches the top of the reservoir.*
- *Highly conductive vugular or solution-enhanced networks in the subsurface act as recipients for mud losses. Any wellbore pressure above pore pressure is likely to result in total losses.*

2.7. Nomenclature.

a	=	empirical factor to calibrate image-derived permeability to field permeability, N
C_0	=	unconfined compressive strength, m/Lt^2, MPa

E	=	Young's modulus, m/Lt^2, MPa
F_a	=	fracture aperture from image logs, L, m
F_d	=	fracture density from image logs, L^{-1}, m^{-1}
g	=	gravitational acceleration, L/t^2, ms^{-2}
h	=	vertical depth, L, ft [m]
K_f	=	fracture permeability, L^2, Darcies
K_I	=	stress intensity factor, $m/L^{-1/2}t^2$, $MPa.m^{1/2}$
K_{IC}	=	fracture toughness, $m/L^{-1/2}t^2$, $MPa.m^{1/2}$
LOT	=	leakoff test
N_Φ	=	triaxial stress factor, N
P_f	=	fluid pressure in crack or fracture, m/Lt^2, MPa
P_p	=	pore pressure, m/Lt^2, MPa
P_w	=	wellbore pressure, m/Lt^2, MPa
S_0	=	cohesion, m/Lt^2, MPa
UCS	=	unconfined (uniaxial) compressive strength, m/Lt^2, MPa
w_w	=	fracture width, L, m
w	=	mechanical aperture, L, m
$\bar{w}$	=	hydraulic aperture, L, m
w_o	=	initial hydraulic aperture, L,m
X_f	=	fracture half-length, L, m
$XLOT$	=	extended leakoff test
α	=	Biot parameter, N
α_T	=	coefficient of thermal expansion, T^{-1}, $^\circ C^{-1}$
γ_h,	=	horizontal reservoir stress path, N
ΔT	=	increase in temperature, T, $^\circ C$
$\Delta\sigma_v$	=	change in total vertical stress, m/Lt^2, MPa
ε_h	=	horizontal tectonic strain in minimum horizontal stress direction, N
ε_H	=	horizontal tectonic strain in maximum horizontal stress direction, N
θ	=	angle between principal stresses on Mohr's circle, degrees
μ	=	coefficient of internal friction in the rock, N
ν	=	Poisson's ratio, N
ρ_b	=	bulk density, m/L^3, g/cc [kg/m^3)
ρ_f	=	density of formation fluid, m/L^3, g/cc [kg/m^3)
σ_θ	=	tangential (or hoop) stress at wellbore wall, m/Lt^2, Mpa
σ_1	=	major principal stress, m/Lt^2, MPa
σ_2	=	intermediate principal stress, m/Lt^2, MPa
σ_3	=	minor principal stress, m/Lt^2, MPa
σ_a	=	axial stress at wellbore wall, m/Lt^2, Mpa
σ_n	=	normal stress, m/Lt^2, Mpa
σ_h	=	minimum horizontal stress, m/Lt^2, MPa
σ_H	=	maximum horizontal stress, m/Lt^2, MPa
σ_v	=	vertical stress, m/Lt^2, MPa
σ_r	=	radial stress at wellbore wall, m/Lt^2, Mpa
σ_{xx}	=	generalized normal stress in x-direction, m/Lt^2, MPa
σ_{yy}	=	generalized normal stress in y-direction, m/Lt^2, MPa
σ_{zz}	=	generalized normal stress in z-direction, m/Lt^2, MPa
τ	=	shear stress, m/Lt^2, MPa
ϕ	=	angle of internal friction, degrees

2.8. References.

Addis, M. A. 1997. Reservoir Depletion and Its Effect on Wellbore Stability Evaluation, *Intl. J. Rock Mech. & Min. Sci.* 34 (3–4): 4.e1–4.e17. https://doi.org/10.1016/S1365-1609(97)00238-4.

Bratton, T. 2020. SPWLA NoW: Rethinking Hydraulic Fracturing—Based on Wellbore Images and Geomechanical Modelling, YouTube video produced by SPWLA International, https://www.youtube.com/watch?v=GeyJTgVLD_s&feature=youtu.be (accessed 27 January 2021).

Cosgrove, J., 2015. The association of folds and fractures and the link between folding, fracturing and fluid flow during the evolution of a fold-thrust belt: A brief Review, in *Geological Society London Special Publications* May 2015, https://doi.org/10.1144/SP421.11.

Dashti, R., Bagheri, M. B., and Shahzad, U. 2009. Fracture Characterizing and Modeling of a Porous Fractured Carbonate Reservoir, SW Iran. Paper presented at the SPE/EAGE Reservoir Characterization and Simulation Conference, Abu Dhabi, UAE, 19–21 October. SPE-125329-MS. https://doi.org/10.2118/125329-MS.

Dyke, C. G., Wu, B., and Milton-Tayler, D. 1995. Advances in Characterising Natural Fracture Permeability from Mud Log Data. *SPE Form Eval* 10 (03): 160–166. SPE-25022-PA. https://doi.org/10.2118/25022-PA.

Economides, M. J. and Nolte, K. G. eds. 2000. *Reservoir Stimulation*, third edition. London, England: John Wiley & Sons.

Esterhuizen, G. S., Dolinar, D. R., and Iannacchione, A. T. 2008. Field Observations and Numerical Studies of Horizontal Stress Effects on Roof Stability in U.S. Limestone Mines, National Institute for Occupational Safety and Heath, *J. of the Southern African Inst of Min Metall* 108 (06): 345–352.

Fisher, K. and Warpinski, N. 2012. Hydraulic-Fracture-Height Growth: Real Data. *SPE Prod & Oper* 27 (01): 8–19. SPE-145949-PA. https://doi.org/10.2118/145949-PA.

Fossen, H. 2012. *Structural Geology* – 3rd printing. Cambridge University Press.

Funatsu, T., Shimizu, N., Kuruppu, M. et al. 2015. Evaluation of Mode I Fracture Toughness Assisted by the Numerical Determination of K-Resistance. *Rock Mech Rock & Eng* 48: 143–157. https://doi.org/10.1007/s00603-014-0550-8.

Hauser, M. R. 2020. Evaluating an Integrated Physical Model for Borehole-Leakoff Pressures. *SPE J.* (2020): 1–21. https://doi.org/10.2118/204453-PA.

Heidbach, O., Rajabi, M., Cui X. et al. 2018. The World Stress Map, database release 2016: Crustal Stress Pattern Across Scales, *Tectonophysics* 744: 484–498. http://doi.org/10.1016/j.tecto.2018.07.007.

Huffman, A. R. and Bowers, G. L. 2003. *Pressure Regimes in Sedimentary Basins and Their Prediction*, AAPG Memoir 76 (January). https://doi.org/10.1306/M76870.

Jadoon, I. A. K., Bhatti, K. M., Siddiqui, F. I. et al. 2005. Subsurface Fracture Analysis in Carbonate Reservoirs: Kohat/Potwar Plateau, North Pakistan. Paper presented at the SPE/PAPG Annual Technical Conference, Islamabad, Pakistan, 28–29 November. SPE-111051-MS. https://doi.org/10.2118/111051-MS.

Jaeger, J. C., Cook, N. G. W., and Zimmerman, R. W. 2007. *Fundamentals of Rock Mechanics,* fourth edition. Hoboken, New Jersey: Blackwell Publishing Ltd.

Kingsley, R. 2018. *Analysis of Fracture Patterns Around the Mena Salt Diapir (Northern Spain), and the Implications for Fluid Flow*, project submitted in part fulfilment for MSc in Integrated Petroleum Science, University of Aberdeen, Aberdeen, Scotland (August 2018).

Kresse, O. and Weng, X. 2013. Hydraulic Fracturing in Formations with Permeable Natural Fractures. In *Effective and Sustainable Hydraulic Fracturing,* ed. R. Jeffrey, J. McLennan, and A. Bunger, London, UK: IntechOpen. https://doi.org/10.5772/56446.

Laubach, S. E., Lander, R. H., Criscenti, L. J. et al. 2019. The Role of Chemistry in Fracture Pattern Development and Opportunities to Advance Interpretations of Geological Materials, *Reviews of Geophysics* **57** (03): 1065–111. https://doi.org/10.1029/2019RG000671.

Laubach, S. E., Olson, J. E., Eichhubl, P. et al. 2010. Natural Fractures From the Perspective of Diagenesis. *CSEG Recorder* **35** (7): 26–31.

Li, B., Guttormsen, J., Hoi, T. V. et al. 2004. Characterizing Permeability for the Fractured Basement Reservoirs. Paper presented at the SPE Asia Pacific Oil and Gas Conference and Exhibition, Perth, Australia, 18–20 October. SPE-88478-MS. https://doi.org/10.2118/88478-MS.

Majidi, R., Miska, S. Z., Yu, M. et al. 2010. Quantitative Analysis of Mud Losses in Naturally Fractured Reservoirs: The Effect of Rheology. *SPE Drill & Compl* **25** (04): 509–517. SPE-114130-PA. https://doi.org/10.2118/114130-PA.

Mohaghegh, S. D. 2013. A Critical View of Current State of Reservoir Modeling of Shale Assets. Paper presented at the SPE Eastern Regional Meeting, Pittsburgh, Pennsylvania, USA, 20–22 August. SPE-165713-MS. https://doi.org/10.2118/165713-MS.

Nelson, R. 2001. *Geologic Analysis of Naturally Fractured Reservoirs,* second edition. Houston, Texas: Gulf Professional Publishing.

Olson, J. E., Laubach, S. E, and Lander, R. H. 2009. Natural Fracture Characterization in Tight Gas Sandstones: Integrating Mechanics and Diagenesis, *AAPG Bulletin* **93** (11): 1535–1549. http://doi.org/10.1306/08110909100.

Piedrahita, J., and Aguilera, R. A. 2017. Petrophysical Dual-Porosity Model for Evaluation of Secondary Mineralization and Tortuosity in Naturally Fractured Reservoirs. *SPE Res Eval & Eng* **20** (02): 304–316. SPE-180242-PA. https://doi.org/10.2118/180242-PA.

Platunov, A., Martynov, M., Nikolaev, M. et al. 2013. Assessment of Geomechanical Concept for Natural and Manmade Fractures in Bazhenov and Tyumenskoe Formations Using an Example of Study in Em-Yoga Field Krasnoleninsky Arch West Siberia. Paper presented at the SPE Arctic and Extreme Environments Technical Conference and Exhibition, Moscow, Russia, 15–17 October. SPE-166914-MS. https://doi.org/10.2118/166914-MS.

Russell, K. A., Cockburn, C., McLure, R. et al. 2005. Improved Drilling Performance in Troublesome Environment. *SPE Drill & Compl* **20** (03): 162–167. SPE-90373-PA. https://doi.org/10.2118/90373-PA.

Russell, K. A., Ayan, C., Hart, N. et al. 2006. Predicting and Preventing Wellbore Instability: Tullich Field Development, North Sea. *SPE Drill & Compl* **21** (01): 12–22. SPE-84269-PA. https://doi.org/10.2118/84269-PA.

Sanderson, D. J. *and* Nixon, C. W. 2015. The Use of Topology in Fracture Network Characterization. *J. Struct. Geol.* **72** (March): 55–66. https://doi.org/10.1016/j.jsg.2015.01.005.

Sayers, C. M. 2010. *Geophysics Under Stress: Geomechanical Applications of Seismic and Borehole Acoustic Waves.* Distinguished Instructor Short Course. Tulsa, Oklahoma: Society of Exploration Geophysicists. https://doi.org/10.1190/1.9781560802129.fm.

Shutterstock. 2021. *Karst, Limestone Erosion Near Dubrovnik, Croatia.* Stock photo ID: 1847832940. https://www.shutterstock.com/image-photo/karst-limestone-erosion-near-dubrovnik-croatia-1847832940 (accessed 2 February 2021).

Siren, T., 2012. *Fracture Toughness Properties of Rocks in Olkiluoto: Laboratory Measurements 2008–2009.* Olkiluoto, Finland: Posiva Oy (May).

Sorkhabi, R. 2014. Fracture, Fracture Everywhere—Part 1, *GEO ExPro* **11** (03). https://www.geoexpro.com/articles/2014/08/fracture-fracture-everywhere-part-i.

Stearns D. W. 1964. Macrofracture patterns on Teton anticline N.W. Montana. In *Transactions of the American Geophysical Union,* Vol. 45: 107. Washington, D.C.: American Geophysical Union.

Suarez-Rivera, R., Behrmann, L., Green, S. et al. 2013a. Defining Three Regions of Hydraulic Fracture Connectivity, in Unconventional Reservoirs, Help Designing Completions with Improved Long-Term Productivity. Paper presented at the SPE Annual Technical Conference and Exhibition, New Orleans, Louisiana, USA, 30 September–2 October. SPE-166505-MS. https://doi.org/10.2118/166505-MS.

Suarez-Rivera, R., Burghardt, J., Stanchits, S. et al. 2013b. Understanding the Effect of Rock Fabric on Fracture Complexity for Improving Completion Design and Well Performance. Paper presented at the International Petroleum Technology Conference, Beijing, China, 26–28 March. IPTC-17018-MS. https://doi.org/10.2523/IPTC-17018-MS.

Warpinski, N. R., Clark, J. A., Schmidt, R. A. et al. 1982. Laboratory Investigation on the -Effect of In-Situ Stresses on Hydraulic Fracture Containment. *SPE J.* **22** (03): 333–340. SPE-9834-PA. https://doi.org/10.2118/9834-PA.

Zhang, X. and Koutsabeloulis, N. 2010. Estimate of Permeability of Fracture Corridors/Networks: from Data Acquisition to Reservoir Simulations. Paper presented at the International Oil and Gas Conference and Exhibition in China, Beijing, China, 8–10 June. SPE-131218-MS. https://doi.org/10.2118/131218-MS.

3. Lost Circulation Mechanisms

There are a number of ways in which whole drilling fluid or cement slurry can leave the wellbore and give rise to partial returns or lost circulation. For some of these ways, the conditions for lost circulation are simple, for others, less so.

The figures in this chapter illustrate the loss of drilling fluid; similar pictures could be drawn for cement slurry.

3.1. Leakage Into Intact Formations. If the drilling fluid or cement has poor fluid loss characteristics, the whole fluid, including its particle content, can invade the formation. The rate of invasion will be higher when the formation is very permeable and/or the overbalance is high; for example, shallow unconsolidated beds or depleted reservoirs. Flow into intact formations, as shown in **Fig. 3.1**, can occur when wellbore pressure is higher than formation pressure, and the pore throats are so large that the fluid cannot form a stable low-permeability filtercake. If it becomes an operational problem, solutions could include reduction of bottomhole pressure, improvement of the fluid loss characteristics of the fluid, or reduction of the pumping rate to reduce erosion of any filtercake.

3.2. Vugular Formations. Carbonate formations frequently contain vugs (pores and pore networks that are significantly larger than the grain size of the rock) and even larger cavities (**Fig. 3.2**; for examples, see Tiab and Donaldson 2016). These are collectively called secondary porosity and are commonly formed by dissolution of the rock, dolomitization, or fracturing. These spaces are big enough that whole drilling

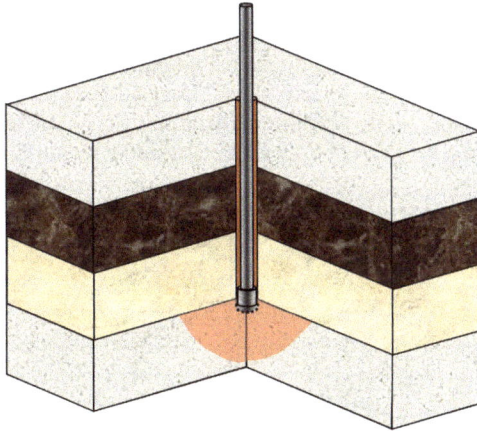

Fig. 3.1—Leakage of whole drilling mud into an intact formation. As the bit penetrates and advances through the formation, it is likely that the loss rate will initially increase sharply then stabilize.

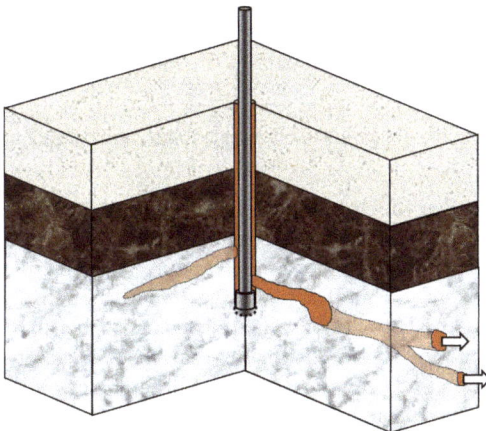

Fig. 3.2—Leakage of whole drilling fluid into cavities in a carbonate formation. The bit first encountered the cavity on the left, which is not connected to a network; after the cavity has been pressurized by drilling fluid, no further losses occur. The bit then drilled into the cavity on the right, which is connected and continues to accept large volumes of drilling fluid.

fluid or cement slurry can flow freely into them if wellbore pressure is greater than formation pressure (Dupriest 2011). This can also happen if the bit penetrates a geological fault in a strong rock, where the roughness of the fault surface has not been smoothed out by shearing and channels for flow still exist.

If the vug, cavity, or fault that the bit encounters is isolated, the initial very large loss rate will drop to zero. If it is connected to a network of cavities, however, rapid losses will continue, until (if there is no intervention) the wellbore fluid level drops to the point where the pressure at the vug equals the formation pressure. This, of course, can lead to serious pressure-control problems elsewhere in the wellbore (Dupriest 2011).

3.3. Natural Fractures. It is very likely—in any well—that the bit will encounter natural fractures in the rock. If a fracture is open at the wellbore (in other words, uncemented and not tightly closed by the prevailing stress in the formation), whole mud or cement slurry might flow into it, as illustrated in **Fig. 3.3**. It is common for several sets of fractures, with different orientations, to be present in a formation, and these may form a network of flow channels leading to substantial loss of fluid.

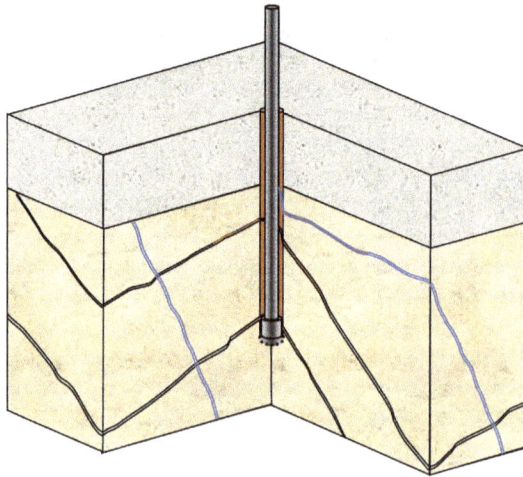

Fig. 3.3—Leakage of whole mud into two sets (black and blue) of natural fractures.

The conditions under which natural fractures can cause serious lost circulation are less simple than for the previous mechanisms. Wellbore pressure must again be higher than the initial fluid pressure in the fractures, which is formation pressure. The aperture or opening of the fracture must be big enough to accept the whole wellbore fluid, including particles of weighting agents such as barite. If the aperture of the fracture at the wellbore wall is smaller than the particles in the fluid, the base fluid will flow into the fracture and particles will filter out on the fracture mouth, forming filtercake and reducing the fluid flow rate.

Some fractures have a more-or-less fixed aperture. For others, though, the aperture may depend on the effective normal stress perpendicular to the plane of the fracture (see Chapter 2). This, in turn, depends on the orientation of the fracture, with respect to the wellbore axis and the stress state in the formation, and on the pressure of the fluid. If filtercake is blocking the fracture mouth, as discussed in the previous paragraph, it may stretch to accommodate changes in fracture aperture, or it may break and allow whole fluid into the fracture. Finally, as fluid flows into a fracture and increases the fluid pressure within it, the fracture aperture might increase and allow greater flow rates.

If fracture apertures do increase, cuttings and other particles from the fluid will flow in and may become trapped. This means that the fractures may not fully close down again when the wellbore pressure is reduced. It is also possible that small-scale shearing of the fracture systems may occur as the pressure within them increases; this would again allow only partial closure of the fractures. So, it is quite possible that short excursions of bottomhole pressure (e.g., surge pressure during trips, connections or casing running) cause a permanent pathway for lost circulation and problems during future operations.

To sum up, it may be difficult to decide in advance whether the presence of natural fractures in a formation is going to lead to serious lost circulation problems. More information is needed on the nature of the fractures, and this can be acquired in a number of ways (see Chapter 4).

If offset well data or experience suggests that natural fractures might open up under the influence of excessive bottomhole pressure, then measures to reduce equivalent circulating density (ECD) during drilling and other operations should be considered. These include changes to the drilling fluid, spacer or cement rheology, limits on the downwards speed of the drillstring or casing, limits on rate of penetration to avoid loading up the annulus with cuttings, and changes to the design of stabilizers and bottomhole assembly. If a quantitative limit on the bottomhole pressure has been established, constant monitoring using a downhole annular-pressure-while-drilling sensor can guide the drilling team (Hutchinson and Rezmer-Cooper 1998).

Finally, lost circulation is often associated with exiting from salt bodies. The state of stress just outside the salt may be significantly different to that inside the salt, which can lead to unexpected fluctuations in the fracture gradient. The rock under an advancing front or tongue of salt may also be very heavily fractured, forming something like a fault breccia or rubble zone. It may be necessary to reduce mud weight and pump pills of lost circulation material, and the sharpness of the transition from salt to native rock means that it is important to know when the boundary will be reached, and to ensure the rig is ready to deal with it.

3.4. Induced Fractures. Recent success with wellbore strengthening methods (sometimes called stress caging) has led to a growth in understanding of induced fractures. These are fractures that do not exist until the drillbit penetrates the formation in question but are initiated from the bit face or wellbore wall by high wellbore pressure. They may or may not lead to problems while drilling or cementing. In fact, if they can be detected by logging instruments during or after drilling, they can give valuable information on the stresses in the formation (Rezmer-Cooper et al 2000). Lost circulation caused by induced fractures can be a limiting factor during infill drilling or drilling through depleted layers. Fracture formation is partly controlled by the minimum principal stress in the formation, which is often decreased by depletion.

The geometry of induced fractures can be complicated, especially when the wellbore is inclined with respect to the principal stress directions in the formation (i.e., in most situations, not vertical) or if the formation is anisotropic. The fractures may not run continuously along the wellbore wall and may be at an angle to the axis of the well. These features can be seen in image logs during or after drilling, but their interpretation takes time, and time may be in short supply during a lost circulation event. It is, however, quite feasible to predict the possibility of induced fractures, to assess the risk they pose and to put plans and materials in place to reduce that risk. Close cooperation with a geomechanics engineer is needed for this.

Fig. 3.4 shows some possible geometries for induced fractures. In general, as they propagate away from an inclined wellbore, they are likely to merge into a single fracture perpendicular to the direction of the minimum principal stress, but the details of how this happens are not well-understood. It is well-established, though, that the pressure for initiating a fracture can be higher or lower than the pressure for propagating it (which is the minimum principal stress). If initiation pressure is higher, exceeding it with static bottomhole pressure or a surge from drillstring or casing motion can lead to unstable fracture growth and lost circulation.

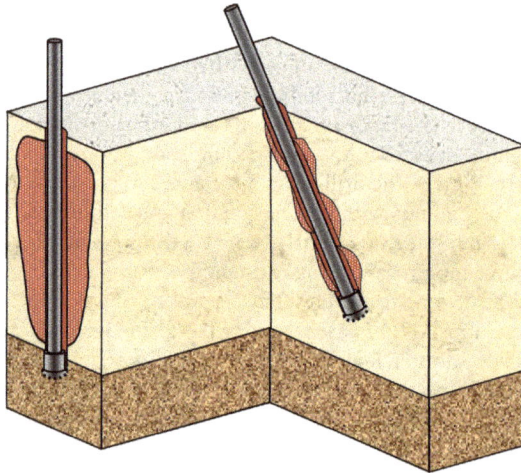

Fig. 3.4—Possible configurations of induced fractures, from (left) a vertical well where a continuous axial fracture forms on each side of the wellbore and (right) an inclined wellbore where discontinuous inclined fractures form (note that these are purely diagrammatic).

Evidently, the conditions under which induced fractures cause *severe* lost circulation are not straightforward:

- Well pressure exceeds fracture initiation pressure and fracture propagation pressure

  Or

- Well pressure exceeds fracture propagation pressure and there is a defect such as an open natural fracture intersecting the wellbore (so that wellbore pressure can be conducted away from the stress concentration around the well)

  And

- Wellbore fluid pressure can penetrate to the tip of the fracture

The second condition is important because the widths of induced fractures are comparable to the sizes of particles in the drilling fluid or cement slurry, and because we expect the width to decrease further away from the wellbore. If there is a suitable particle-size distribution in the fluid, the solids can create a filtercake on the mouth of the opening fracture and keep it isolated from wellbore pressure, as discussed for natural fractures in Section 3.3. They can also flow into the fracture and form a bridge where the width decreases. This bridge can hinder or block fluid flow further into the fracture and the full wellbore pressure may not reach the tip of the fracture. It is well-established from studies of hydraulic fracturing that this partial penetration increases the wellbore pressure needed to propagate the fracture. This is the basis of design of drilling fluids for wellbore strengthening (or stress caging, as it is sometimes known). Engineered fluids for wellbore strengthening will be discussed in Chapter 5.

The discussion of filtercake effects in this section relates mainly to water-based fluids. In nonaqueous fluids, such as invert emulsion oil-based muds, the filtercake is much thinner and weaker and is much less likely to bridge the mouth of a growing fracture.

3.5. Borehole Breathing. When fracture initiation pressure is less than propagation pressure, high wellbore pressure can generate fractures that do not propagate. This can lead to wellbore breathing, where the well accepts drilling fluid when the pumps are on and flows back when they are switched off. The fractures are inflated (their widths are increased) by the increased bottomhole pressure caused by frictional pressure drop in the annulus. When the pumps are off, the frictional pressure drop is reduced and the fractures close again, rejecting fluid into the annulus. It is important that the drilling team is aware of this possibility, so that the outflow is not mistaken for a kick. There are diagnostic signs of borehole breathing in both surface (Dyke et al. 1995) and downhole (Bratton et al. 2001) data. Once again, close cooperation with a geomechanics engineer is recommended.

3.6. Depleted Reservoirs. When reservoir pressure is reduced through production, the minimum principal stress in the formation can decrease (Addis 1997). This can be qualitatively predicted using a simple geomechanics model, but quantitative prediction is more difficult and may require full-scale 3D mechanical modeling of the field (Talreja et al. 2019).

Reduced minimum stress means reduced fracture gradient and greater risk of lost circulation during drilling and cementing. In, for example, stacked sand-shale sequences, past production may decrease fracture gradient in the sands and make infill drilling or access to deeper reservoirs very difficult. The alternation of low fracture gradient in multiple depleted sands with high mud weight requirements in the intervening shales, means that it is not possible to find economical or practical casing schemes.

This situation led to the development of drilling fluids engineered to allow drilling with mud weights and bottomhole pressures above the fracture gradient in the sands. One of the first approaches to this was the so-called "Stress Cage" technique, in which hard, sized particles were included in a drilling fluid with very low fluid loss (Aston et al. 2004; Alberty and McLean 2004). The intention was to trap the hard particles in the fractures close to the wellbore wall, and so, generate additional hoop stress and prevent further fracturing. Another approach used combinations of more-resilient sized particles to form a barrier to fluid flow within the fracture, inhibiting the transmission of wellbore pressure to the fracture tip (van Oort et al. 2009). A third approach used a high-fluid-loss mud to pack solids into the fracture, building width and "fracture closure stress" (Dupriest 2005). Whichever technique or model is preferred, these have all had considerable success in drilling through depleted reservoirs. A drawback common to them all is that they need some permeability in the formation—it is much more difficult to make them work in rocks with very low permeability such as shale. This is because they all rely on flow through the walls of the fracture into the formation pore space, either to decrease the pressure behind a particle bridge, or to allow leakoff from the drilling fluid.

Engineered fluids for drilling above the fracture gradient will be further discussed in Chapter 5.

3.7. Lost Circulation During Casing and Cementing. During casing running and cementing operations, tubulars and viscous fluids are being introduced into the wellbore. We should therefore expect changes in the bottomhole pressure, just as in drilling operations, and increased risk of lost circulation. The loss of fluid from the wellbore might stop after a temporary surge in bottomhole pressure, or it might create fractures and cause a permanent route for losses. For example, a study of lost circulation (Therond et al. 2018) showed that 90% of losses were initiated either during circulation before cementing or during casing running.

The reasons for this may be physical or operational/organizational. Cleaning the wellbore after reaching total depth (TD) for a section means flushing out cuttings and cavings. This may need high drilling-fluid flow rates and consequent high ECD. If cuttings or cavings settle on top of the drilling collars, they may restrict the annulus with the same effect. When casing is run into hole, the narrow clearances can lead to large annular pressure drops. Operational reasons might include the feeling that TD has been achieved, so the section or well is finished and less care needs to be taken: the need for speed in operations, no longer limited by rate of penetration; changes of crew between drilling, casing running, and cementing; and less-detailed monitoring of parameters such as downhole pressure and vertical speed of the drill or casing string.

On the positive side, cement slurry has a higher solids content than drilling fluid. If formation of filtercake over fracture mouths is a significant factor in preventing lost circulation, slurry will do this more effectively.

Summary
Whole mud or cement slurry can be lost from the wellbore in four distinct ways:

- *Flow into formations with very large pore size*
- *Flow into secondary porosity, such as vugs and caverns*
- *Flow into natural, pre-existing fracture networks*
- *Flow into fractures induced by the pressure in the wellbore*

All these mechanisms require that wellbore pressure be higher than formation pressure (this is, of course, good drilling practice anyway). Natural and induced fractures, though, respond to temporary increases in wellbore pressure by opening and/or propagating. This means that careful monitoring and control of downhole pressure during drilling, casing running, and cementing can reduce the risk of lost circulation from these mechanisms.

3.8. References.

Addis, M. A. 1997. The Stress-Depletion Response Of Reservoirs. Paper presented at the SPE Annual Technical Conference and Exhibition, San Antonio, Texas, 5–8 October. SPE-38720-MS. https://doi.org/10.2118/38720-MS.

Alberty, M. W. and McLean, M. R. 2004. A Physical Model for Stress Cages. Paper presented at the SPE Annual Technical Conference and Exhibition, Houston, Texas, 26–29 September. SPE-90493-MS. https://doi.org/10.2118/90493-MS.

Aston, M. S., Alberty, M. W., McLean, M. R. et al. 2004. Drilling Fluids for Wellbore Strengthening. Paper presented at the IADC/SPE Drilling Conference, Dallas, Texas, 2–4 March. SPE-87130-MS. https://doi.org/10.2118/87130-MS.

Bratton, T. R., Rezmer-Cooper, I. M., Desroches, J. et al. 2001. How to Diagnose Drilling Induced Fractures in Wells Drilled with Oil-Based Muds with Real-Time Resistivity and Pressure Measurements. Paper presented at the SPE/IADC Drilling Conference, Amsterdam, The Netherlands, 27 February–1 March. SPE-67742-MS. https://doi.org/10.2118/67742-MS.

Dupriest, F. E. 2005. Fracture Closure Stress (FCS) and Lost Returns Practices. Paper presented at the SPE/IADC Drilling Conference, Amsterdam, The Netherlands, 23–25 February. SPE-92192-MS. https://doi.org/10.2118/92192-MS.

Dupriest, F. E. 2011. Kick Mechanisms and Unique Well Control Practices in Deep-water Vugular Carbonates. Paper presented at the International Petroleum Technology Conference, Bangkok, Thailand, 15–17 November. IPTC-14423-MS. https://doi.org/10.2523/IPTC-14423-MS.

Dyke, C. G., Wu, B. and Milton-Tayler, D. 1995. Advances in Characterising Natural Fracture Permeability From Mud Log Data, *SPE Form Eval* **10** (03): 160–166. SPE-25022-PA. https://doi.org/10.2118/25022-PA.

Hutchinson, M. and Rezmer-Cooper, I. 1998. Using Downhole Annular Pressure Measurements to Anticipate Drilling Problems. Paper presented at the SPE Annual Technical Conference and Exhibition, New Orleans, Louisiana, 27–30 September. SPE-49114-MS. https://doi.org/10.2118/49114-MS.

Rezmer-Cooper, I., Bratton, T., and Krabbe, H. 2000. The Use of Resistivity-at-the-Bit Images and Annular Pressure While Drilling in Preventing Drilling Problems. Paper presented at the IADC/SPE Drilling Conference, New Orleans, Louisiana, 23–25 February. SPE-59225-MS. https://doi.org/10.2118/59225-MS.

Talreja, R., Bahuguna, S., and Havelia, K. 2019. Merits of 3D Numerical Over 1D Analytical Geomechanics Solutions for a Complex Subsurface. Paper presented at the SPE Oil and Gas India Conference and Exhibition, Mumbai, India, 9–11 April. SPE-194693-MS. https://doi.org/10.2118/194693-MS.

Therond, E., Taoutaou, S., James, S. G. et al. 2018. Understanding Lost Circulation While Cementing: Field Study and Laboratory Research. *SPE Drill & Compl* **33** (01): 77–86. SPE-184673-PA. https://doi.org/10.2118/184673-PA.

Tiab, D. and Donaldson, E. C. 2016. *Petrophysics: Theory and Practice of Measuring Reservoir Rock and Fluid Transport Properties,* fourth edition. Houston, Texas: Gulf Professional Publishing.

van Oort, E., Friedheim, J. E., Pierce, T. et al. 2009. Avoiding Losses in Depleted and Weak Zones by Constantly Strengthening Wellbores. Paper presented at the SPE Annual Technical Conference and Exhibition, New Orleans, Louisiana, October 2009. https://doi.org/10.2118/125093-MS.

4. Measurement, Prediction, and Diagnosis

4.1. Data—Interpretation and Action. This chapter deals with collection of data about lost circulation events and interpreting them to help choose which actions to take. In general, the more data that can be obtained about a problem, the easier and more appropriate the solution. In drilling and cementing operations, however, the timing of the data or information enters the balance.

If a well is being drilled in the first stages of a large project, then extensive wireline logging after drilling and modeling with the acquired data are valuable. A picture of the behavior of the field on a large scale can be vital in understanding which formations and drilling practices are likely to lead to problems on future wells, such as lost circulation, assisting the planning, and reducing costs of individual wells and the entire project.

For more immediate problems with the well being drilled, however, data are needed that are acquired in real time, quickly interpreted, and lead to timely actions. This approach has been successful in the control of wellbore instability (Bradford et al. 2000) and should be applied to lost circulation as well. There are many new sources of data—surface and downhole—that can be relevant to lost circulation. These often, of course, mean extra planning and cost, but this should always be carefully balanced against the risk of significant unplanned costs and downtime because of lost circulation, or because of, for example, wellbore instability caused by an overcautious approach to wellbore pressure management. Another valuable aspect of real-time data collection is to calibrate and refine drilling practice; for example, using annular pressure measurements to *know* what effect trip speed has on equivalent circulating density (ECD), rather than to *predict* it, and to modify tripping practice (slower or faster) accordingly.

In addition to the advances in data collection and ever-improving telemetry rates from downhole tools, methods of dealing with the data have undergone some radical changes in the past 20 years or so, with great strides in machine learning, data analytics, Bayesian processing, and other approaches. These can be linked to automated systems on the rig or may simply present processed information to the drilling team. These data-processing systems have some great advantages: they don't need to sleep, change shifts, or deal with broken mud pumps. They can incorporate experience from many other wells and drilling teams, as well as complex calculations of non-Newtonian mud flow. They are unlikely to replace the skill of experienced drillers in the near future, but they can be a valuable and untiring assistant.

4.2. Detecting and Measuring the Losses. When total losses occur, it may be difficult to assess the rate of loss, and there may be limited value in the number anyway.

The loss rate will be the sum of the loss from the annulus and the pumping rate. If there is an annular pressure sensor in the bottomhole assembly (BHA) and the mud pumps are still running (so that mud-pulse telemetry of the pressure data is still possible), the decrease with time of bottomhole pressure will give a guide to the rate of loss from the annulus. This guide will be lost when the top of the mud is in a highly deviated or horizontal section, and there are likely to be more pressing priorities on the rig anyway.

For smaller loss rates, the value can be used in diagnosis of the mechanism, and so can help in choosing a remedy. For many years, flow into the well has been measured from the stroke rate of the mud pumps, with corrections for pump efficiency and fluid compressibility. Flow out has often been measured with a paddle flowmeter on the mud-return line. These are relatively coarse measurements but have the advantage of being on almost every rig; no additional installation is required. The difference between flow out and flow in is frequently monitored (using these sensors or others) to give early warning of kicks; it can, of course, also be used for losses. Problems arise if this monitoring is automated to make an alarm that does not need constant human attention; for example, tripping pipe in or out of the well leads to extra or missing outflow and can trigger false alarms. There are approaches to overcome this using multiple data channels from the operations of the rig, not just pump rate and paddle position, but also block height, hook load, rate of penetration, and others. These data can be fed to a sophisticated signal-processing and pattern-recognition system using Bayesian methods to discover what is happening on the rig and to produce a sensible and reliable indicator of kicks and losses (Aldred et al. 2008; Erge et al. 2017). Advantages of this multiple-data-channel approach are its relative insensitivity to missing data (e.g., because of a broken connection) and its flexibility to accept other sensors (such as a flowmeter rather than a pump-stroke counter).

Verga et al. (2000) uses electromagnetic flowmeters and finds them "very effective to monitor mud loss rates to detect the presence of fracture or fracture network during drilling." Beda et al. (2001) also uses electromagnetic flowmeters and other rig data channels in a "time domain" analysis of losses, allowing quantification of mud losses and their relation to the productivity index measured later on the wells.

More recently, Chiniwala et al. (2018) discusses a similar approach called *advanced flow analysis*, again using many rig data channels and mud flow data to generate a log of loss and fracture characteristics. This work is more aimed at reservoir characterization than drilling, and so, uses data from mud-gas analysis by gas chromatography and cuttings analysis by x-ray fluorescence, but their collection of mud flow and rig data allows them to detect the mud losses in real time and to classify them on the basis of mechanism. **Fig. 4.1** shows an example of the output of the advanced flow analysis approach.

In the examples shown by Chiniwala et al. (2018), the mud flows are again measured with electromagnetic flowmeters, but they mention the possibility of using Coriolis flowmeters instead. Coriolis flowmeters are bulkier than electromagnetic flowmeters, and more expensive, but are more accurate and, importantly, they work with nonconducting fluids such as oil-based muds. Electromagnetic flowmeters can only work with conductive fluids such as water-based muds. Coriolis meters are influenced by the presence of gas in the mud, but the errors are small for small gas fractions; if the gas fraction is high, it will be obvious in other ways.

Both electromagnetic and Coriolis flowmeters are unaffected by the rheology of the drilling mud.

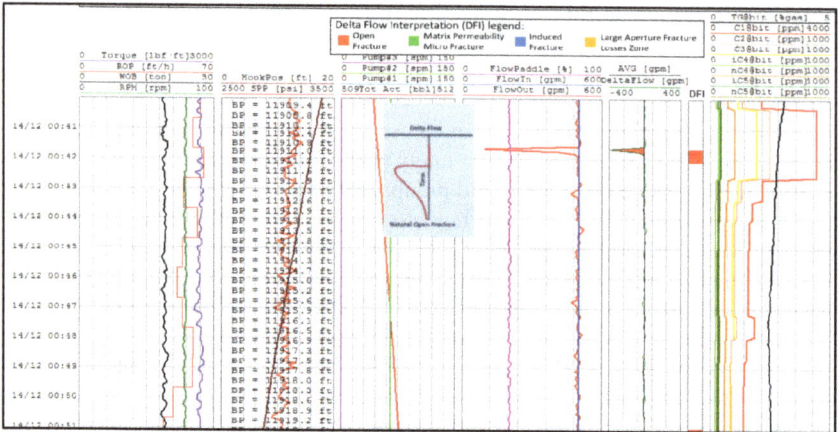

Fig. 4.1—An open natural fracture detected from its fingerprint on an Advanced Flow Analysis log. This figure reprinted from Chinawala et al. (2018), with permission from URTec, whose permission is required for further use.

4.3. Location and Identification of the Loss Zone. Correctly identifying the position of the lost circulation or thief zone is critical for the proper choice and placement of a lost circulation treatment. Previous drilling records and local experience can give an approximate idea of when a lost circulation zone might be expected. Drilling behavior, changes in the rate of fluid loss, formation boundaries (seen from the mud log or from changes in rate of penetration) and many logging techniques can help in narrowing down the location.

Lost circulation is often assumed to begin at bit depth or the last casing shoe, but this is not necessarily the case. If the cause is a geological feature, such as an open fracture, a very high-permeability formation, or a vug, then losses will begin or increase as the bit reaches it. If, however, the cause is an induced fracture or opening up of a closed natural fracture, this could occur in the openhole section above the bit, driven by an increase in the annulus pressure, which is driven in turn by a change in drilling conditions. For example, suppose the bit has drilled successfully through an interval with a low minimum stress, but the annulus pressure was not high enough to initiate losses. If the bit then enters a zone of weak rock, the rate of penetration increases, and the annulus will begin to load up with more cuttings. This could lead to an increase in annulus pressure or ECD that is enough to initiate losses in the higher interval.

Indicators for losses at bit depth include the following:

- They occur while drilling ahead under steady conditions, rather than when tripping, starting the mud pumps, or changing their rate.
- A drilling break (i.e., a sudden increase in rate of penetration, including free fall of the bit) as the bit penetrates a highly fractured zone, a vug, or other geological features

- They coincide with the occurrence of bit balling, whose symptoms are decreasing rate of penetration and weight-on-bit and a substantial decrease in torque. The buildup of solids around the bit increases the mud pressure at the bit face and can lead to fracturing there

Indicators for losses above the bit include the following:

- They occur as the drillstring is moving down rapidly, as during a trip in or a connection
- They coincide with a reduction in the volume of cuttings and cavings seen at surface, indicating a buildup of solids in the annulus or a pack-off around the BHA
- They occur as the rate of penetration or hole instability increases, loading up the annulus with cuttings or cavings

Several logging methods are available for locating the zones where losses are occurring.

4.3.1. Temperature, Flow, and Tracer Measurements. If fluid is being lost to the formation, generally it will be flowing from above the depth of loss. So, after a period of stabilization, the temperature of the well fluid below the loss zone will be close to that of the formation, but the fluid above will be cooler. Running a temperature log past the loss zone will reveal its location as a step change in temperature. A hot wire sensor can also be used to detect the flow itself. Both these methods take some time, with fluid being lost all the time, so they may require large volumes of fluid.

The loss zone can also be determined by direct logging of the wellbore flow, using venturi or spinner flow monitors conveyed on wireline tools. Again, fluid must be being lost for the flow to be significant, so large volumes may be needed.

Radioactive surveys to locate the thief zone require making two gamma ray surveys. A base log is run first. Then, a pill of mud containing radioactive material is pumped down the hole. A new log is run, and high concentrations of radioactive material will be noted at the loss zone. This method provides accurate data for locating the point of loss but requires expensive equipment and additional deliberate loss of mud to obtain the desired data. It also involves the use of radioactive materials, which is forbidden in many environments.

4.3.2. Resistivity Logs. Resistivity tools measure the electrical resistance of the formation around the wellbore. This can be used for formation evaluation; for example, to determine the saturation of hydrocarbon and its mobility. They can also be used in some circumstances to detect fractures in the rock around the wellbore. There are different forms of resistivity tool, and they are generally available during drilling as part of the logging-while-drilling suite, as through-bit logging tools, or as wireline tools after drilling (and to some extent, after casing and cementing).

Laterologs (Petrowiki 2020a) measure formation resistivity by monitoring the voltage required to feed alternating electrical current through the borehole fluid into the formation. There are many variants working at different frequencies with multiple electrodes or arrays of electrodes, many depths of investigation, and azimuthal sensitivity (seeing different responses at different orientations around the wellbore axis). All, however, require conductive fluid in the wellbore.

Induction tools (Petrowiki 2020b) use coils to induce currents in the formation. This means that they can work with nonconductive wellbore fluids. Again, there are many variants with different frequencies, multiple coils or arrays, azimuthal sensitivity, and various depths of investigation (away from the wellbore axis and ahead of the bit).

Propagation tools send electromagnetic waves with relatively high frequencies through the formation and measure their attenuation and phase shift. The two measured parameters, at different transmitter/receiver spacings and different frequencies, can be interpreted to give a comprehensive picture of resistivity around the wellbore.

Electrical resistivity tools can also be used to form images of the wellbore wall. This will be discussed in the next section.

Resistivity tools can be used to detect features in the formation that lead to lost circulation, such as fractures and invasion. When wellbore fluids enter the formation, the local resistivity may be changed. This could, for example, be with conductive water-based drilling fluid invading a resistive hydrocarbon-bearing formation, or with nonconductive oil-based fluid entering fractures and displacing conductive formation fluid. The different depths of investigation of resistivity logging, and the possibility of running repeat logs, allow detection of these changes with very good depth control.

Shazly and Tarrabees (2013) show examples of the estimation of fracture apertures in the hard Nubian Sandstone using the Dual Laterolog resistivity tool. The deep and shallow responses of the tool were modeled with a finite-element approach, giving ranges of possible apertures for each of the fractures detected (**Fig. 4.2**). These

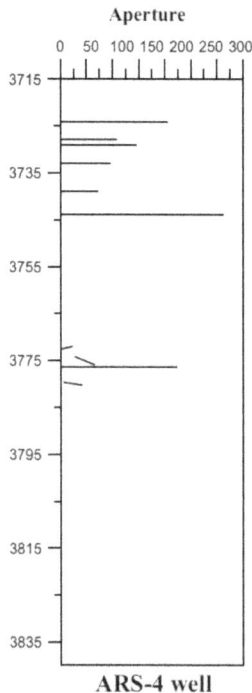

Fig. 4.2—Computed fracture aperture ranges (in microns) vs. depth (in meters) in Nubian Sandstone, Well ARS-4 in the Abu Rudeis-Sidri Field, Egypt; from Shazly and Tarrabee (2013).

ranges are sometimes very broad (e.g., 0–270 microns), sometimes more limited (25–75 microns), but together with other measurements, can help to identify and characterize loss zones.

Repeat or time-lapse resistivity logging can be very useful in the right circumstances, when a generation of fractures, or invasion of drilling fluid into them, changes the resistances and current paths around the wellbore. Repeat logging with tools carried on wireline normally needs a repeat run, but when the tools are in the drillstring, the usual actions of drilling, short and long trips, working the pipe, for example, can mean that the resistivity sensors pass the same point in the wellbore many times. This offers an excellent opportunity to view the initiation or progression of fractures and to modify drilling tactics accordingly.

Fig. 4.3 shows an example from Zhang et al. (2020) of time-lapse resistivity logging in the context of lost circulation and its treatment. The data are from a vertical well drilled using synthetic invert-emulsion mud, with lost circulation between 25,450 and 25,620 ft. The left-hand track shows the rate of penetration (red) and gamma ray (black). The five tracks to the right show resistivity measured during repeat passes with a logging-while-drilling tool. The purple regions show where measured formation resistivity has been increased because drilling-induced fractures have been invaded by the nonconductive drilling fluid. The repeated measurements allow tracking the effects of different treatments for lost circulation.

4.3.3. Resistivity Imaging Tools. Imaging resistivity tools use pads pressed against the wellbore wall, containing arrays of closely spaced electrodes, or buttons. Each button independently directs current into the formation and back to a return electrode. The close spacing of the buttons allows construction of a high-resolution (better than 1 cm) map of the resistivity of the wellbore wall, which can show up geological features such as formation dip, vugs, and natural and induced fractures. Imaging tools were initially only available on wireline conveyance and in conductive drilling fluid. Many advances were made in the numbers of pads and buttons and the coverage of the wellbore wall. Imaging is now available in nonconductive drilling fluids, such as oil-based mud, and also while drilling, with button arrays mounted on stabilizers as part of a logging-while-drilling assembly. **Fig. 4.4** shows an example of the evolution of the quality of imaging-while-drilling tools from Shrivastava et al. (2019).

4.3.4. What the Driller or Cementing Engineer Needs to Know. Logging tools and techniques, particularly those conveyed by the drillstring, have made tremendous advances in the past 30 years or so. Making repeated logging runs to assess fracture invasion, or acquiring an image of the open hole, no longer means tripping out and running wireline tools in hole. Great advances have been made in the data rate that can be transmitted by mud-pulse telemetry, so it is now completely practical to collect this data, turn it into useful information, and act on it on a timescale relevant to drilling and cementing operations. The additional costs of running these tools needs to be set against the potential costs (and lost time) of running logs after drilling to assess a problem, plus the costs associated with solving a problem that could have been avoided with the help of timely information.

4.4. Diagnostics. In the past, the choice of treatment for lost circulation would have been based on experience, decision trees, availability of materials, and trial and error.

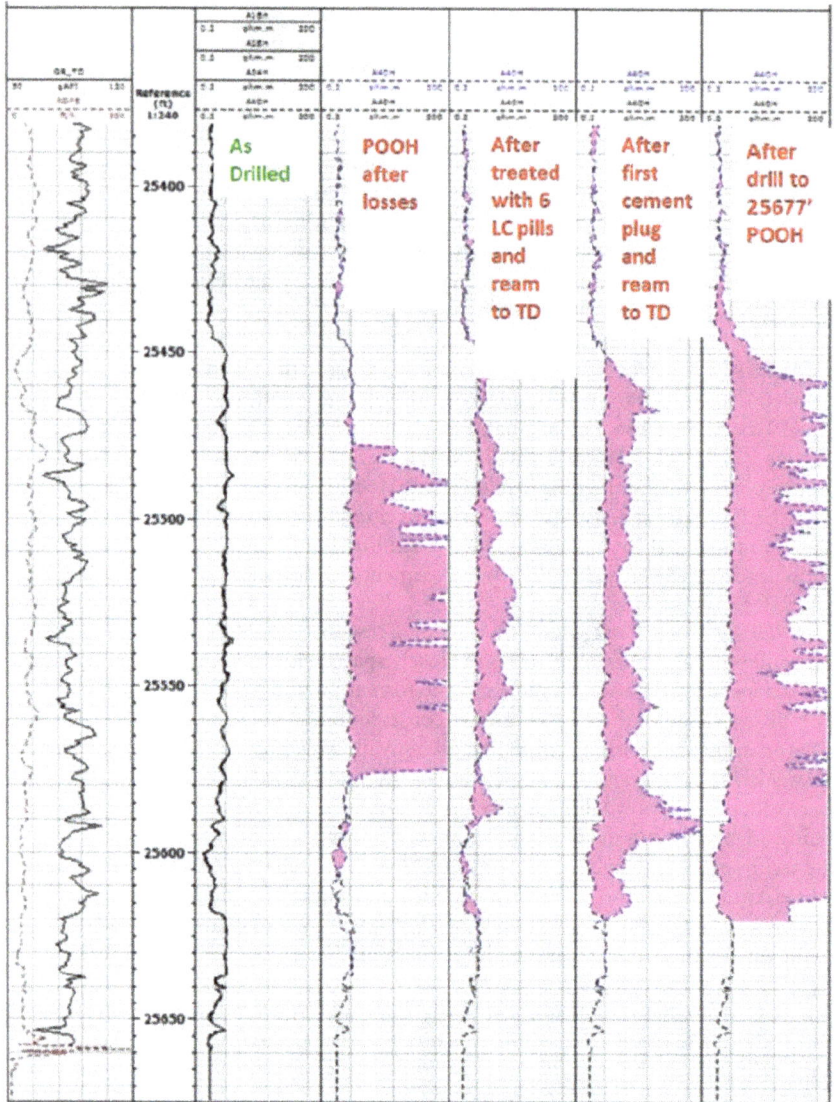

As Drilled

POOH after losses

After treated with 6 LC pills and ream to TD

After first cement plug and ream to TD

After drill to 25677' POOH

Fig. 4.3—Repeated resistivity logs (using a logging-while-drilling tool) showing invasion of fractures by nonconductive drilling fluid; from Zhang et al. (2020).

Now, as treatments have become more specific and directed, it is important to know as much as possible about the mechanisms of lost circulation active in a particular well along with some specific data, such as fracture width.

An early paper by Dyke et al. (1995) is still a good guide to initial diagnosis of lost circulation. **Fig. 4.5**, which is based on Dyke et al.'s Fig. 2, shows how the mud tank level is expected to react to different loss mechanisms.

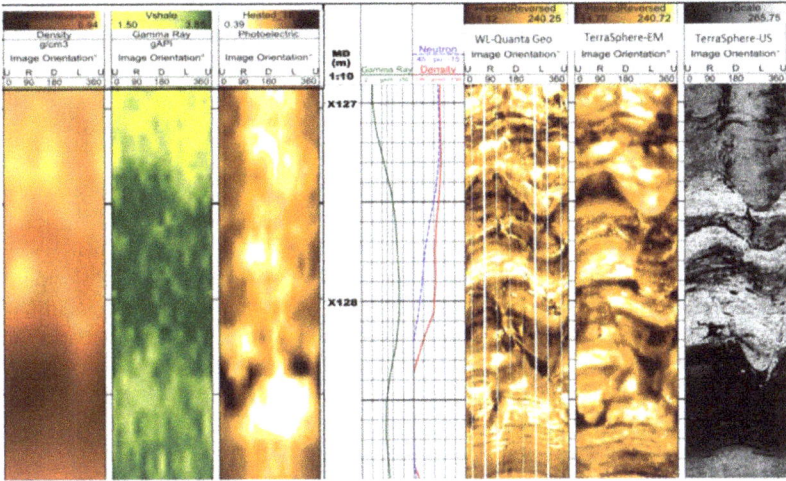

Fig. 4.4—Various types of image log showing 2 m of a fractured formation. The three panels on the left show images from legacy-while-drilling tools: oriented density, oriented gamma-ray, and photoelectric factor. Although these were very useful when introduced, limited detail is visible. The three right-hand panels show (left to right) a resistivity image from a wireline-conveyed pad tool; a resistivity image formed with an array of button electrodes on a logging-while-drilling tool; and an image from the same LWD tool but using an ultrasonic sensor. The dual-physics approach allows a very complete characterization of the fractures in real time; from Shrivastava et al. (2019, fig. 2).

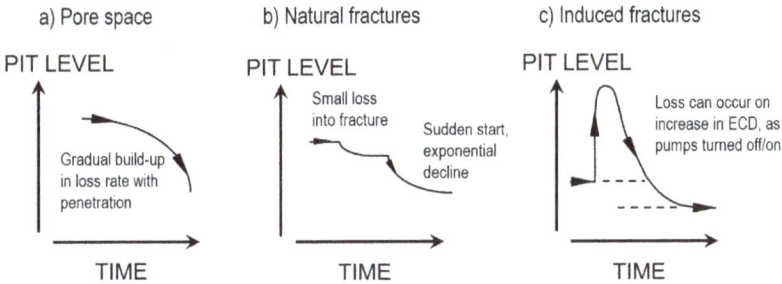

Fig. 4.5—Expected behavior of the fluid level in the mud tank in response to different types of lost circulation events; after Dyke et al. (1995), Fig. 2. If the losses are of whole mud into a highly permeable formation (a), the rate of loss will increase as the bit makes progress and more and more formation is exposed, so the rate at which the tank level drops will increase. If the losses are to natural fractures (b), the losses will initially be rapid, but slow down as mud particles begin to block the fracture and the viscosity of the mud generates additional resistance to flow (if the natural fracture begins to open under the influence of the mud pressure, this decline may not occur). If the losses are to induced fractures (c), wellbore breathing can occur; if the pumps are switched off at a connection, ECD will drop and mud will be ejected from the fractures as they close; this will be seen at surface as an increasing tank level. When the pumps are turned on again, this is reversed, and the fractures may grow. Losses to vugs and cavities are not shown here—the tank level would simply decrease linearly as the mud pumps empty the tank.

Dyke et al. (1995) go on to discuss in detail how different kinds of fractures are blocked, and how more-detailed diagnosis of fracture characteristics can inform the production strategy.

There are many other sources of data available on the rig, on a timescale relevant to preventing, diagnosing, and curing lost circulation. Sonic logging, using excitation and detection of Stoneley waves, can give the location of fractures intersecting the wellbore, their width, and whether or not they are sealed or open. Until recently, this information was only available on wireline, limiting its usefulness for the current well, but can now be done while drilling using advanced LWD sonic tools.

The use of annular pressure while drilling data in preventing lost circulation will be discussed in Chapter 5, but it can also be used in diagnosis. There are many tools available from service companies that measure annulus pressure with high accuracy and sampling rate and can transmit the value to surface via mud-pulse telemetry. Bratton et al. (2001) show how the shape of the measured ECD curve as the mud pumps are turned on or off is diagnostic of the presence of fractures in the open hole. This interpretation can give early warning during drilling that fractures are growing from the wellbore and may become unstable and lead to losses. **Fig. 4.6** shows the difference between the two types of responses.

As mentioned in Section 4.2, reliable mud flow-in and flow-out meters can be fitted and can quantify the rates of loss. In the case of losses into fractures, these values can be used with fluid mechanics models to estimate some parameters of the fracture and improve the choice of methods for curing the losses. The models can be complex, because of the non-Newtonian rheology of drilling fluids and the ability of fractures to open up as the pressure within them increases. Lietard et al. (1996, 1999) demonstrates the use of a model for mud loss rate into fractures and discusses the wisdom of using lost circulation materials in naturally fractured reservoirs during drilling and cementing. Verga et al. (2000) uses an analytical model to calculate fracture widths from mud loss rates and validates the results using image logs. More recently, there have been many studies examining different aspects of the rate of mud loss into fractures (including Lavrov and Tronvoll 2004; Tempone and Lavrov 2008; Majidi et al. 2008; Huang et al. 2011; and Bychina et al. 2017).

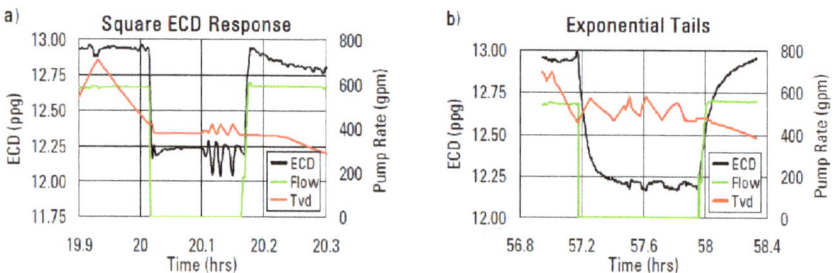

Fig. 4.6—Response of the measured ECD to a pump-off/pump-on sequence in an interval with (a) unfractured rock, and (b) fractured rock. The ECD curves also show the significant response of the pressure to small movements of the drillstring (red Tvd curve); after Bratton et al. (2001).

Table 4.1—Data types for comprehensive lost circulation analysis, after Zhang et al. (2020).

Rate of loss information in offset wells	Temperature profile survey
Cumulative losses by depth	Seismic interpretations
Resistivity logs	Fault conductivity analysis
Pore pressure and fracture gradient predictions	Bit depth location
Formation pressure integrity tests	Loss and gain behavior
Formation porosity, permeability, and pore throat size	Drilling fluid properties including water phase salinity, background lost circulation material
Lost circulation (pill formulation and performance	Casing and liner pressure tests
Borehole image logs	Fracture width analysis tool
Lithology determination and gamma ray logs	Torque, drag, and rate of penetration
Tectonic regime / in-situ stress state	Well trajectory information
Annular pressure	Wireline logs for casing integrity

Zhang et al. (2020) discusses a very comprehensive approach to diagnose and characterize losses, primarily looking at natural and induced fractures. They use 22 types of input data, listed in **Table 4.1,** and include estimates of fracture width. The approach allows the selection of an optimum treatment, either during well planning or during drilling operations. They reach a number of interesting conclusions, very relevant to understanding losses into fractures. One in particular is as follows:

"For conductive natural fractures/faults, no significant loss is expected if the estimated fracture width is less than 200 μm, and normal [lost circulation materials] LCMs cannot work effectively if the estimated fracture is wider than 1000 μm."

In some circumstances, knowing the locations and characteristics of fractures is enough to allow the design of a lost circulation treatment; for example, the size distribution and concentration of particles of loss prevention material or the required strength and gelling time of a chemical treatment. Some workers have gone further, however, and developed integrated approaches, taking data from many sources and using artificial intelligence or data-driven techniques to design a treatment. Some of these approaches will be discussed in the next section.

4.5. Real Time Data and Decision-Making Software. Lost circulation is a complex problem with a variety of causes and mechanisms, occurring at any time during the drilling and cementing processes, and with potentially severe consequences. Experience is a good guide in such situations but not all well planners and rig crews are equally experienced, so a number of workers have designed software systems to help with decision making. These can be based on physics or on machine-learning techniques [perhaps using artificial neural networks (ANNs)] or on statistical and signal-processing approaches such as Bayesian networks. ANN methods have to be trained or calibrated on initial datasets that include input data and outcomes.

Hou et al. (2020) developed an ANN approach for extreme high-pressure/high-temperature (HPHT) wells in the Yingqiong Basin in South China. Lost circulation was a persistent problem in the Huangliu Formation at approximately 4000 m depth. They developed an ANN system to predict six severities of lost circulation (including no losses). They used a wide range of input parameters from 50 wells: mud weight, mud yield point, solids content, plastic viscosity, pump rate, rotation speed, rate of penetration, weight on bit, standpipe pressure, bit nozzle area, measured depth, lithology, fracture and pore pressure gradients (from the drilling plan) and rock strength. The ANN was trained on these data and was able to make a good prediction of lost circulation severity within the training set of 50 wells. It is not clear whether the predictions have extended to wells outside the training set, but the authors are confident, on the basis of the convergence behavior of the ANN, that it would perform well. This type of system could be adapted by training on well data local to a particular area, then used as a prediction system during well planning and in real time during well operations. It could also be extended to cementing operations.

Ren et al. (2020) use a physics-based approach for losses into fractures, modeling flow into a deforming fracture, with leakoff through the fracture faces. The model generates families of type curves, and data from the rig is objectively matched to the type curves using a feature-extraction algorithm called the Hough Transformation. This allows construction of a map of loss type (e.g., natural fractures and caverns). The coordinates of the map are a dimensionless parameter from the Hough Transformation and the difference between wellbore and formation pressure, and the parameter plotted on the map can be fracture width (to help with treatment selection) or loss mechanism.

Fig. 4.7 shows these two maps.

Ren et al. (2020) also discuss data acquisition for lost circulation analysis and make the point that the accuracy of Coriolis flowmeters makes this feasible.

Ahmed et al. (2020) compares two machine-learning techniques, an ANN and a functional network. They train the algorithms on one well and use them to predict lost circulation events in another well. The input parameters for the network training

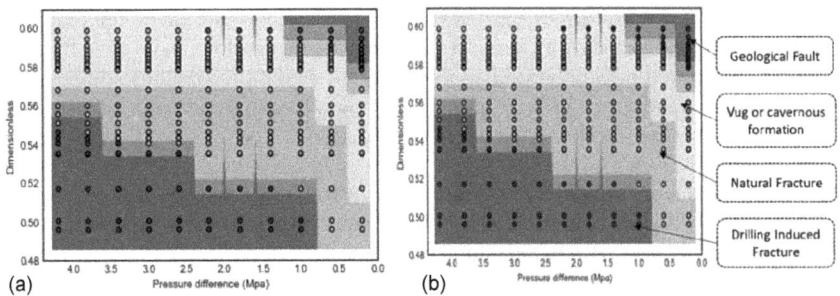

Fig. 4.7—Color-coded maps of (a) fracture width and (b) lost circulation mechanism plotted against dimensionless parameters derived from type curves fitted to mud loss data and pressure difference between wellbore and formation; after Ren et al. (2020).

are pump flow rate, rate of penetration, rotary speed, standpipe pressure, drilling torque, and weight on bit. The outcomes are simply the occurrence of lost circulation. Both techniques worked well in predicting problems in the second, or test, well, with low false-alarm rates.

These integrated approaches are still in their infancy, and it may be some time before they gain acceptance. They potentially have many advantages, as discussed in Section 4.1. The speed of progress in development of machine-learning methods, and their widespread use in many other areas of technology and of everyday life, indicates that they will play a major role in dealing with lost circulation problems in future.

Summary

Action on lost circulation can be taken in the planning stage or while operations are in progress, but efficient action always requires some data and information on the mechanisms of loss. The types of data and the extent of the processing or interpretation for these two cases are different; planning data can use dense data and detailed interpretation, but during operations, the data available may be liable to be interrupted, and the time available for interpretation shorter. There has been a lot of progress recently on computer-aided methods that can automate data processing and interpretation, and these are expected to be more widely used in future.

There have also been great advances in the hardware available to collect data, mainly in the real-time domain. Mud flow sensors are becoming more mature and can give a picture of the losses accurate enough to allow diagnosis of the mechanisms. Images of the borehole are now routinely available while drilling—at high-enough resolution to be very useful in diagnosis—and telemetry has improved to allow transmission to surface of these images without compromising more traditional data requirements, such as well inclination and azimuth.

Lost circulation can have a severe impact on the well and project viability, and if it is a risk factor, the extra costs of data collection for diagnosis should be carefully balanced against the consequences of drilling and cement jobs.

4.6. References.

Ahmed, A., Elkatatny, S., Ali, A. et al. 2020. Prediction of Lost Circulation Zones Using Artificial Neural Network and Functional Network. Paper presented at the Abu Dhabi International Petroleum Exhibition & Conference, Abu Dhabi, UAE, 9–12 November. SPE-203268-MS. https://doi.org/10.2118/203268-MS.

Aldred, W. D., Hutin, R., Luppens, J. C. et al. 2008. Development and Testing of a Rig-Based Quick Event Detection System to Mitigate Drilling Risks. Paper presented at the IADC/SPE Drilling Conference, Orlando, Florida, 4–6 March. SPE-111757-MS. https://doi.org/10.2118/111757-MS.

Beda, G. and Carugo, C. 2001. Use of Mud Microloss Analysis While Drilling to Improve the Formation Evaluation in Fractured Reservoir. Paper presented at the SPE Annual Technical Conference and Exhibition, New Orleans, Louisiana, 30 September–3 October. SPE-71737-MS. https://doi.org/10.2118/71737-MS.

Bradford, I. D. R., Aldred, W. A., Cook, J. M. et al. 2000. When Rock Mechanics Met Drilling: Effective Implementation of Real-Time Wellbore Stability Control.

Paper presented at the IADC/SPE Drilling Conference, New Orleans, Louisiana, 23–25 February. SPE-59121-MS. https://doi.org/10.2118/59121-MS.

Bratton, T. R., Rezmer-Cooper, I. M., Desroches, J. et al. 2001. How to Diagnose Drilling Induced Fractures in Wells Drilled with Oil-Based Muds with Real-Time Resistivity and Pressure Measurements. Paper presented at the SPE/IADC Drilling Conference, Amsterdam, The Netherlands, 27 February–1 March. SPE-67742-MS. https://doi.org/10.2118/67742-MS.

Bychina, M., Thomas, G. M., Khandelwal, R. et al. 2017. A Robust Model to Estimate the Mud Loss into Naturally Fractured Formations. Paper presented at the SPE Annual Technical Conference and Exhibition, San Antonio, Texas, 9–11 October. SPE-187219-MS. https://doi.org/10.2118/187219-MS.

Chiniwala, B., Palakurthi, A. K., Easow, I. et al. 2018. Measurement and Analysis of Wellbore Micro Losses and Rock Properties While Drilling: A Novel Approach to Identification of Fractures in the Osage and Meramec Formations of Anadarko Basin. Paper presented at the SPE/AAPG/SEG Unconventional Resources Technology Conference, Houston, Texas, 23–25 July. URTEC-2896976-MS. https://doi.org/10.15530/URTEC-2018-2896976.

Dyke, C. G., Wu, B., and Milton-Tayler, D. 1995. Advances in Characterising Natural Fracture Permeability From Mud Log Data. *SPE Form Eval* **10** (03): 160–166. SPE-25022-PA. https://doi.org/10.2118/25022-PA.

Erge, O., de Mata Cecilio. I., Dearden, R. et al. 2017. Detection of Influx and Loss of Circulation. International Patent No. WO/2017/059153.

Hou, X., Yang, J., Yin, Q. et al. 2020. Lost Circulation Prediction in South China Sea using Machine Learning and Big Data Technology. Paper presented at the Offshore Technology Conference, Houston, Texas, 4–7 May 2020. OTC-30653-MS. https://doi.org/10.4043/30653-MS.

Huang, J., Griffiths, D. V., and Wong, S. W. 2011. Characterizing Natural-Fracture Permeability from Mud-Loss Data. *SPE J* **16** (01): 111–114. SPE-139592-PA. https://doi.org/10.2118/139592-PA.

Lavrov, A., and Tronvoll, J. 2004. Modeling Mud Loss in Fractured Formations. Paper presented at the Abu Dhabi International Conference and Exhibition, Abu Dhabi, United Arab Emirates, 10–13 October. SPE-88700-MS. https://doi.org/10.2118/88700-MS.

Lietard, O., Unwin, T., Guillot, D. et al. 1996. Fracture Width LWD and Drilling Mud / LCM Selection Guidelines in Naturally Fractured Reservoirs. Paper presented at the European Petroleum Conference, Milan, Italy, 22–24 October. SPE-36832-MS. https://doi.org/10.2118/36832-MS.

Lietard, O., Unwin, T., Guillot, D. et al. 1999. Fracture Width Logging While Drilling and Drilling Mud/Loss-Circulation-Material Selection Guidelines in Naturally Fractured Reservoirs (includes associated papers 75283, 75284, 81590, and 81591). *SPE Drill & Compl* **14** (03): 168–177. SPE-57713-PA. https://doi.org/10.2118/57713-PA.

Majidi, R., Miska, S. Z., Yu, M. et al. 2008. Modeling of Drilling Fluid Losses in Naturally Fractured Formations. Paper presented at the SPE Annual Technical Conference and Exhibition, Denver, Colorado, 21–24 September. SPE-114630-MS. https://doi.org/10.2118/114630-MS.

Petrowiki. 2020a. Electrode Resistivity Devices (24 June 2015 revision), https://petrowiki.spe.org/Electrode_resistivity_devices (accessed 29 December 2020).

Petrowiki. 2020b. Induction Logging (24 June 2015 revision), https://petrowiki.spe. org/Induction_logging (accessed 29 December 2020).

Ren, R., Miska, S. Z., Yu, M. et al. 2020. Physics-Based Data-Driven Approach for Downhole Fracture Inference Using Lost Circulation Data. Paper presented at the 54th U.S. Rock Mechanics/Geomechanics Symposium, physical event cancelled, June 2020. ARMA 20–1564.

Shazly, T. F. and Tarabees, E. 2013. Using of Dual Laterolog to Detect Fracture Parameters For Nubia Sandstone Formation in Rudeis-Sidri Area, Gulf of Suez, Egypt. *Egyptian Journal of Petroleum* **22** (02): 313–319. https://doi.org/10.1016/j. ejpe.2013.08.001.

Shrivastava, C., Maeso, C., Wibowo, V. et al. 2019. Multi-Measurement Logging-While-Drilling Imager: New Enabler for Wide-Scale Comprehensive Geosciences Applications in Oil-Base Mud. Paper presented at the Abu Dhabi International Petroleum Exhibition & Conference, Abu Dhabi, UAE, 11–14 November. SPE-197402-MS. https://doi.org/10.2118/197402-MS.

Tempone P. and Lavrov A. 2008. DEM Modelling of Mud Losses into Single Fractures and Fracture Network. Paper presented at the 12th International Conference of International Association for Computer Methods and Advances in Geomechanics, Goa, India, 1–6 October.

Verga, F. M., Carugo, C., Chelini, V. et al. 2000. Detection and Characterization of Fractures in Naturally Fractured Reservoirs. Paper presented at the SPE Annual Technical Conference and Exhibition, Dallas, Texas, 1–4 October. SPE-63266-MS. https://doi.org/10.2118/63266-MS.

Zhang, J., Sant, R., Majidi, R. et al. 2020. Treatment of Losses Based on Root Cause and Fracture Width Analysis. Paper presented at the SPE Annual Technical Conference and Exhibition, Virtual, 26–29 October 2020. SPE-201317-MS. https://doi.org/10.2118/201317-MS.

5. Preventing and Curing Lost Circulation

5.1. Planning the Well. During the planning phase, well engineers must assess the potential for lost circulation and the risks associated with it. They must plan the casing scheme, drilling fluid program, and cementing design to reduce these risks, balancing them against risks from other sources such as kicks and hole stability, the availability of materials and equipment at the well site, and, of course, the cost of operations. Sometimes severe lost circulation from a high equivalent circulating density (ECD) must be balanced against hole instability while drilling with a lower ECD; it is worth remembering that there are more ways of coping with an unstable hole than with severe lost circulation.

All parts of drilling, casing, and cementing operations need careful planning. For example, in a study performed for offshore wells in four different regions, 24 wells were evaluated. From a total of 72 casing strings, 40 had losses reported at some time during drilling and cementing. The study showed that 90% of the losses were initiated during either circulation before cementing or casing running; only 10% were initiated during the drilling phase and the actual pumping of cement (Therond et al. 2017).

5.2. Drilling and Well Engineering Solutions. In addition to conventional well design engineering techniques, such as drilling fluids selection, ECD management, casing design, bit selection and rate of penetration, to name a few, the well engineer can tackle lost circulation risks with prevention or mitigation measures (or both). The preventative measures are listed below and will be discussed in subsequent sections:

- During drilling and cementing, avoid inducing losses and keep the unfractured rock intact. This can be done with techniques such as the different options for managed pressure drilling (MPD), casing while drilling (CWD), expandable tubulars, and managed pressure cementing (MPC). Monitor bottomhole pressure carefully if there is a risk of lost circulation.
- While running the casing, avoid surge and swab; establish limits for the speed of movement of the casing and make sure they are followed. Surging or swabbing the well will create excess pressure that could induce losses. Use automated fill-up tools to reduce these occurrences.
- Design cementing fluids with controlled densities and viscosity to reduce the friction pressure in the annulus and control the equivalent circulating density, taking into account the importance of hydraulics for good mud removal.

5.2.1. Managed Pressure Drilling (MPD). MPD is defined by the International Association of Drilling Contractors as "an adaptive drilling process used to more precisely control the annular pressure profile throughout the wellbore" (Frink 2006). It should be investigated during the well design process to determine whether it is economically viable. The primary objective of MPD is to obtain a stable wellbore within a narrow operating pore pressure and fracture gradient window, while avoiding any losses or fluid influxes.

During MPD, the top of the annulus is closed by a rotating control device (RCD), which allows the drillstring to be rotated and moved vertically. The mud returns are passed through a variable choke (**Fig. 5.1**). These devices are coupled with automated

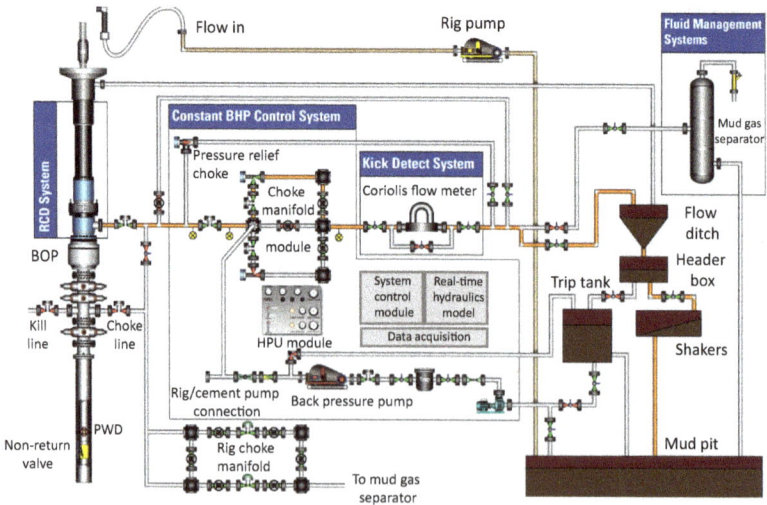

Fig. 5.1—Managed pressure equipment during drilling/cementing; after Wahid et al. (2014).

pumps and flow measurement devices, all controlled by software running a hydraulics model. The objective is to control the annular pressure profile within the wellbore.

MPD clearly requires extra equipment, time, and cost, but this should be compared to the costs associated with severe lost circulation. MPD techniques include constant bottomhole pressure (CBHP), dual-density gradient, and pressurized mud cap drilling.

Constant Bottomhole Pressure (CBHP). Even when standard MPD is used, the bottomhole pressure can vary significantly. In particular, when making a connection, there is no mud flow in the annulus and no friction pressure, so the ECD drops. **Fig. 5.2** illustrates this. If the mud weight window between pore pressure and fracture gradient is narrow, this can mean having to choose between influxes

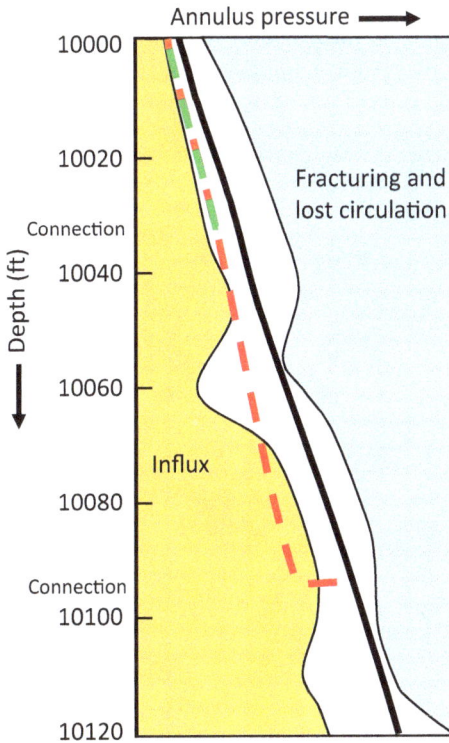

Fig. 5.2—Comparison between conventional and CBHP techniques. For safe drilling without influxes or losses, the ECD should stay at all times in the window between pore pressure (orange) and fracture pressure (blue). During conventional drilling, even with an MPD system in place, the reduced flow rate during connections causes decreases in frictional pressure drop and ECD. The green and red dashed lines illustrate this for two connections spaced apart by a double stand of pipe. The wellbore pressure drops along the entire annulus. If influxes or excessive connection gas are seen, it is difficult to raise the mud density without losing circulation as the pumps are turned on again. Drilling with a continuous flow system eliminates the reductions during connections and allows a constant ECD (black line).

and losses. The CBHP technique uses additional hardware to maintain circulation while making connections, avoiding fluctuations in bottomhole pressure. For example, one system uses a sub with a side-entry valve at the top of each stand of drillpipe. During connections, flow from the mud pumps is diverted from the standpipe into the valve and then into the drillsting, so annular flow and ECD are maintained (Riddoch et al. 2016).

Dual-Density Gradient Drilling. This technique is used when the pore pressure gradient increases rapidly with increasing depth so that neither a static nor a dynamic column of a single-density fluid can be maintained without fracturing the shallower formations. The technique can be used onshore and offshore, especially in deep water environments (Stave 2014).

Offshore, a variety of dual-density gradient methods have been proposed and used. Some methods use pumps at the seafloor or at an intermediate water depth to lift the returning mud to surface and decrease the wellbore pressure. Other methods dilute the returning mud, at the seafloor or above, with unweighted fluid. Onshore, air or nitrogen is injected into the annulus at the casing shoe or elsewhere.

Fig. 5.3 illustrates the use of dual-density gradient drilling in an offshore well.

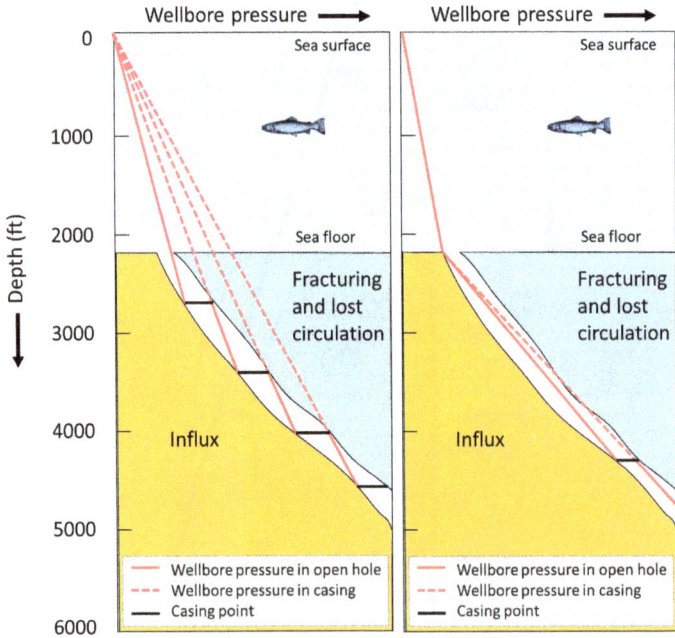

Fig. 5.3—Principle of dual-density gradient drilling technique for offshore drilling. Both diagrams show pore pressure and fracture gradients, mud pressures, and casing points as a function of depth. On the left, mud pressures for conventional drilling rise from zero at the sea surface and as a result, four mud weights and casing points are needed to reach below 5,000 ft. On the right, mud pressures rise from the hydrostatic value at the sea floor, and so, a much more economical casing plan is possible with only one casing point.

Pressurized Mud Cap Drilling (PMCD). The PMCD technique can be used when drilling through fractured and vuggy formations where losses occur at formation pressure. It uses a column of mud in the annulus, which is lighter than that required to balance the formation pressure.

With this technique, drilling is performed using an RCD, with the well shut in at the surface (**Fig. 5.4**). The surface annular pressure is used as an indicator of what is occurring downhole. Sacrificial fluid, usually water, is pumped down the drillstring, and a lighter annular mud is pumped into the annulus. All fluid and cuttings are pumped into the fractures or vugs. Maintaining a full hole with a static column of fluid reduces losses of the expensive drilling fluid (Dipura et al. 2018).

The advantages of PMCD are the reduction of drilling mud cost, better well control, higher rate of penetration (because of reduced overbalance pressure), and reduced nonproductive time.

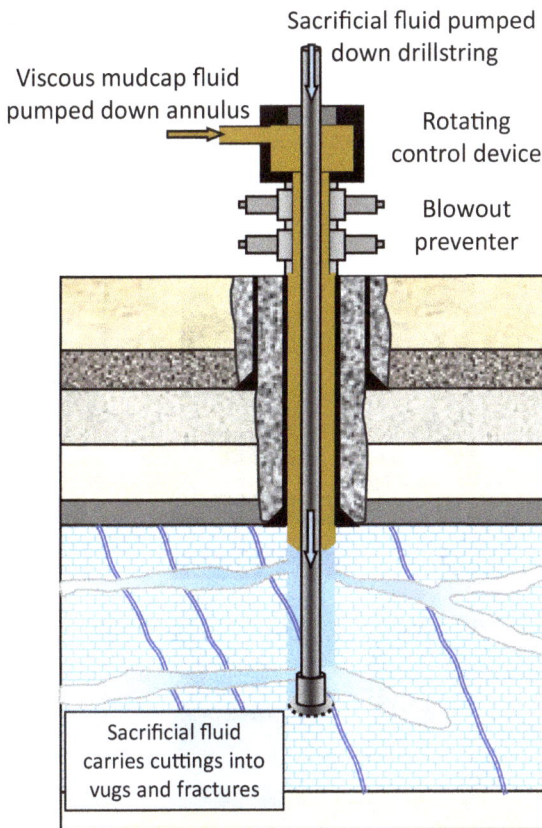

Sacrificial fluid pumped down drillstring

Viscous mudcap fluid pumped down annulus

Rotating control device

Blowout preventer

Sacrificial fluid carries cuttings into vugs and fractures

Fig. 5.4—Basic principle of PMCD. The low-cost sacrificial fluid is pumped down the drillstring and is lost into vugs and fractures, while the mudcap fluid and rotating control device maintain wellbore pressure.

Managed Pressure Cementing (MPC). The first trials using MPC date to 2004 in the North Sea. The cementing operations were engineered to avoid losses by controlling the pressure. This technique has been used successfully in different environments around the world for high pressure/high temperature wells in Malaysia (Wahid et al. 2014), Saudi Arabia (Siddiqi et al. 2016), the North Sea (Bjorkevoll et al. 2008), and the Caspian Area (Rajabi et al. 2012).

The main objective is to place the cement slurry safely in the annulus, controlling the ECD to stay within the specified margins. Lavrov (2016) presents a mathematical model for the flow of slurry within the well and any fracture that is present; in particular, he points out the balance that must be struck between high-yield point and viscosity (reducing propagation and invasion of a fracture) and low-yield point and viscosity (reducing the ECD fluctuations that create and widen the fracture).

The algorithms for design and control of MPC are more complex than for MPD because they deal with several fluids (drilling mud, spacers, and cement slurries) with different properties. Each fluid has its own density and rheology, and they flow together in the drillstring and annulus. The MPC hardware system has to deal with static and dynamic conditions. In static conditions, it must maintain the annular pressure to avoid losses and influx using the backpressure pump. In dynamic conditions, it must place the fluids in the right place, taking into account all the changes in backpressure from different flow rates and the friction pressures they generate. Control is through the automated choke and return line coming from the RCD. The key components of the MPC process are shown in **Fig. 5.5**.

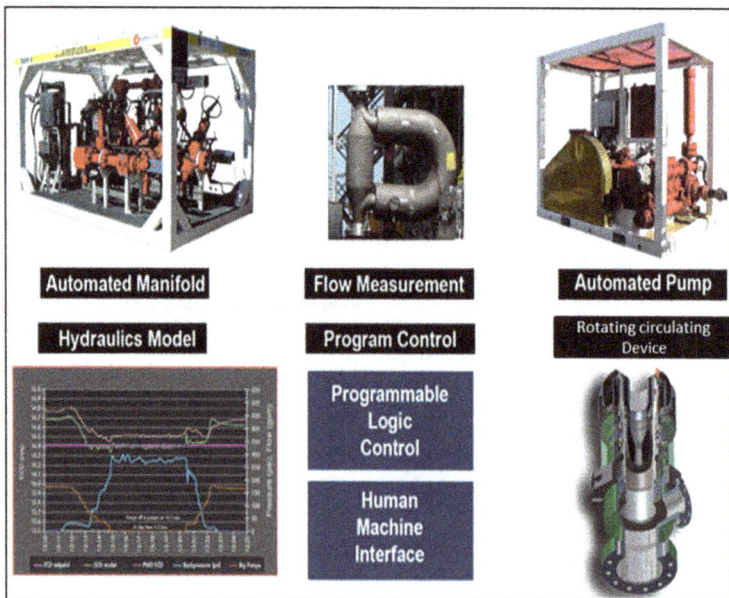

Fig. 5.5—MPD/MPC main components comprising automated manifold and pumps, flow measurement devices, rotating control (or circulating) device, hydraulic model, and programmable logic control systems; from Wahid et al. (2014).

The complexity of the MPC fluids placement, and the impact of the annular pressure drop of several fluids with different properties, requires a rigorous design process using MPD software in conjunction with cementing hydraulic simulators to control ECD.

5.2.2. Casing While Drilling (CWD). The first patent on drilling with casing (Chapman 1913) was filed in 1890, describing a technique of drilling with casing and retrieving the bottomhole assembly (BHA) including the hydraulically expandable drilling bit. The industry, however, had to wait until the early 2000s to witness deployment of suitable technology and to reap its benefits (Houtchens et al. 2007).

During CWD, the well is drilled and cased simultaneously using standard oilfield casing; the drilling assembly—consisting of the drilling bit and an underreamer to enlarge the open hole and allow the casing to pass through—is latched into the bottom of the casing as it is run and retrieved through the casing at the end of the bit run (**Fig. 5.6**). Retrieving the assembly through the casing eliminates borehole damage and allows safer tripping, thus minimizing/eliminating the risk of lost circulation.

The well can be drilled vertically or directionally using a rotary steerable drilling assembly fitted with directional equipment, such as mud motors and measurement-while-drilling (MWD) for trajectory control.

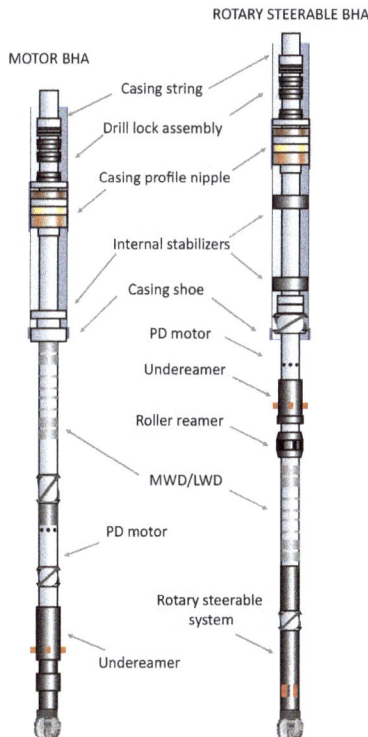

Fig. 5.6—Different BHA configurations for CWD, using a motor or a rotary steerable system.

The profile nipple in Fig. 5.6 is used to land cementing equipment, which is pumped down after the drilling assembly is removed to allow the cement job to take place.

CWD eliminates the critical and risky operation of running the casing, where most induced losses occur. It also allows reduction of losses if they do occur, because the small annular clearance between the casing and the open hole inhibits annular flow. The rotary motion of the casing string also helps, causing smearing or plastering of the cuttings, mud solids, and filtercake onto the wellbore walls, filling up fluid-leakage paths (Fontenot et al. 2003).

CWD is a cost-effective drilling technique that reduces the number of operations procedures by eliminating casing running and reducing/eliminating wellbore instability, stuck pipe, and lost circulation. It also contributes to long-term downhole well integrity by ensuring that the cement job is done properly and that the primary barrier—in this case cement—has been placed, tested, and verified. One of the advantages of CWD downhole geometry is that the difference between the casing outer diameter and the openhole diameter is in the order of 20 mm, instead of approximately 50 mm for the conventional configuration. This has two effects: it helps to plaster the wellbore and so reduce the lost circulation risks, and it helps in placing cement at high velocities and ensuring a good cement coverage (Sánchez et al. 2012).

The pumpdown plugs and floats used for cementing after CWD are critical components and must be designed or selected carefully to meet well integrity requirements.

5.2.3. Expandable Tubulars. Casing off an openhole section normally means that the diameter of the next section must be reduced. If an unexpected problem (such as severe lost circulation or shallow water flows) occurs before the planned end of the section and needs to be put behind casing, this changes the casing plan for the rest of the well and might mean that final production rates are no longer economically viable.

Expandable tubular technology overcomes this problem by running a casing string into the openhole section, then expanding it to the previous casing diameter. The concept emerged in the 1990s, and a variety of companies developed systems for deployment (Cales 2003). The ability to expand casing in situ enables conservation of internal casing diameter in a planned way, which allows larger casing sizes in the reservoir to produce at higher rates than are otherwise practical (Filippov et al. 1999).

The technology has been widely used and, even when not actually used, it can be held in readiness as a contingency measure to de-risk a drilling plan.

There are several variants of expandable tubular technology, but the basic concept of all is to cold work the casing downhole, using a cone or other means to produce permanent outwards deformation of the pipe. Typically, expansions of more than 25% of the diameter of the pipe can be accomplished. The expansion process can be done in two ways: upward expansion applies pressure beneath the cone to propel it through the pipe and deform the metal into its plastic region; downward expansion uses rollers, which travel from top to bottom of the tubular, pushing and rotating at the same time. With upward methods, the casing is cemented then expanded, with downward methods, it is expanded then cemented (**Fig. 5.7**).

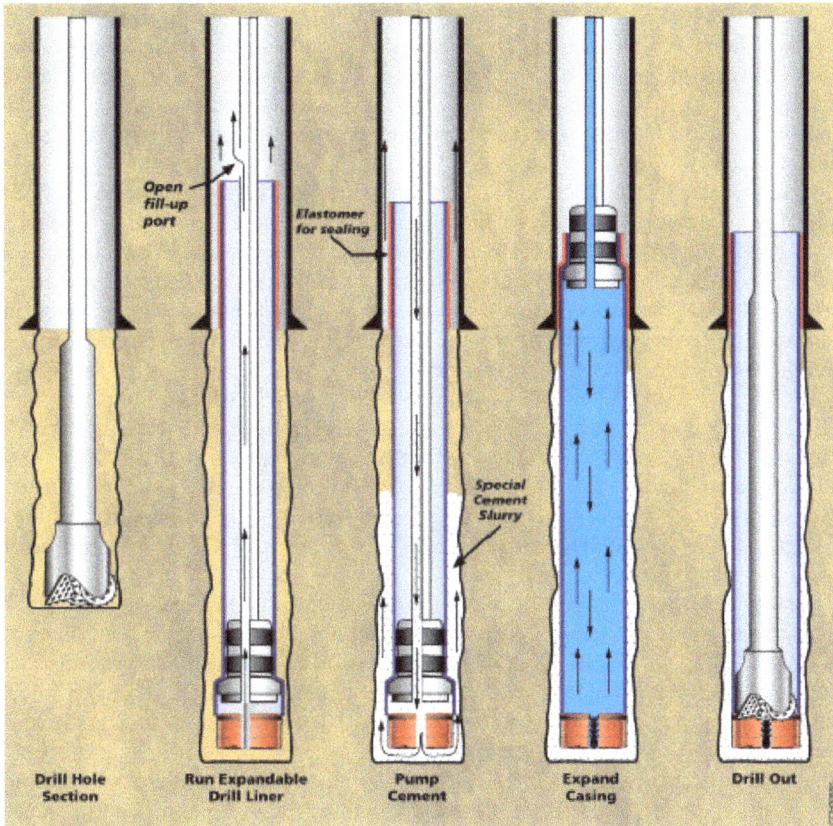

Fig. 5.7—Expandable tubular installation using upward expansion: an example of the process from makeup, including cementing and drillout; from Filippov et al. (1999).

5.2.4. Lost Circulation Materials and Fluids. Hundreds of materials and fluid systems have been proposed and used as lost circulation solutions. They have many different objectives and activation methods. Many aim simply to clog the pathways for fluid flow out of the wellbore, either at the wellbore wall or within the formation; others to form high-viscosity mixtures or gels to slow or inhibit fluid flow; and others to set solid in the wellbore and fluid pathways needing to be drilled out afterwards.

They can be placed or activated simply by pumping and letting the fluid take the materials to the loss zone, or by spotting and squeezing, with or without special equipment in the drillstring, by time or temperature, or by shearing a mixture through the bit nozzles to trigger a chemical reaction. The choice of material, fluid, or placement method depends on the suspected mechanism of lost circulation, the drilling fluid type, and the anticipated temperature and wellbore pressure they must withstand, so diagnosis on the basis of experience and data from offset and current wells is very important. The choice of action also depends, realistically, on what materials and equipment are at hand or are easily available.

Another important criterion is the behavior of the material during production. For example, in many wells (especially in carbonates), the hydrocarbons are expected to be produced from fractures. If fractures in the producing interval are permanently blocked by lost circulation material used during drilling, well productivity will be reduced. Lost circulation materials for production intervals should be either degradable or removable (e.g., with acid).

"Nondesigned" Solid Materials. Over the years, all kinds of solids have been added to the drilling fluid to try to block fluid-loss pathways. These include shredded rope, paper, and plastic film; high-melting-point asphalts, such as gilsonite; expanded perlite; mica; nut shells; and even the paper sacks in which such additives are delivered to the rig. Sometimes these are chosen with a particular size range, perhaps on the basis of local experience, or what is quickly available. Many are cost effective, but there are some limitations/requirements.

Particles can be added to cement slurry to prevent or reduce losses. Taoutaou et al. (2013) showed that cement slurries containing resilient graphitic carbon particles can plug fractures up to 7.2-mm aperture. Graphitic carbon particles are generally considered to be "resilient" if, after applying a compaction pressure of 70 MPa (10,000 psi), the particles expand and recover at least 20% of their original volume. For the present use, resiliency of at least 80% is preferred. A US patent in this area was published by Taoutaou et al. (2013).

If it is hoped that these will block crack or vug openings at the wellbore wall, it is important to control drilling practices so that any buildups of the materials are not removed by the bit, stabilizers, or swab pressures. Of the additives mentioned, only asphalt has any tendency to stick to the wellbore wall, and then, only if it has softened at bottomhole temperatures. Sharp-edged stabilizers, back reaming, and frequent wiper trips should be avoided.

It is also vital to consider the clearances within the components of the drilling assembly. These materials are all designed to prevent fluid flow through openings and will certainly tend to block openings in MWD turbines/sirens, rotary-steerable system valves, filters, and other critical parts of a modern drilling system. Most of these systems now have specifications of what will or will not block them. If the proposed lost circulation material is outside these specifications, it is possible to include a bypass sub in the drillstring to deliver the materials to the annulus without going through the equipment behind the bit.

Fibrous Materials. Some of the materials mentioned in the previous section were fibrous. Everyday experience shows us that fibers are good at blocking openings. There have been a number of approaches in recent years, however, to optimize their performance—for example, to design fibers or blends that block more effectively at low concentrations or to increase the strength of the packed fiber mass.

Fibers aimed at bridging over fractures can be tailored by their length and bending stiffness, so they can form a mat over openings and then trap particles from the drilling fluid or cement to form an impermeable layer. Pumpability is very important; the fibers must be chosen so that they do not clog the rig pumps, nor be broken up by passage through the valves. Wettability is also important, because this influences whether the fibers will clump or disperse in particular fluids. These tailored treatments are aimed at particular sizes of openings, so it is important to have a procedure established for choosing the correct formulation (Pasteris et al. 2014).

Taoutaou et al. (2006) discuss many of these issues in the context of a fiber additive for losses during drilling and cementing in the depleted Brent Field of the North Sea. They also address several of the other ways to mitigate lost circulation problems, including drilling fluid design, and drilling/cementing practices.

Fibers of different kinds (mineral and organic) can also be blended into a high-fluid-loss pill, pumped down, then squeezed to dewater it and form a dense mass, for example, within vugs or cavities. Once again, compatibility with the drilling fluid and tools and with the downhole environment are important (for examples, see Sanders et al. 2010; Sanders and Houston 2014).

Gels. Another way to slow down or stop flow into fractures and bigger apertures without blocking the drilling tools is to pump or squeeze materials (solids-free or with a small particle size) that form a gel or set to a solid in the annulus. This is the basis of a treatment that has been used for many years, usually called a gunk plug. Unhydrated bentonite is mixed with diesel and pumped down to the suspected loss zone. When it encounters water (either in the drilling fluid or from the formation), it thickens and gels rapidly to slow the losses. Cement may be added to the mixture for a stronger set, the pill may be squeezed into the formation, and if a nonaqueous drilling fluid is being used, the bentonite in diesel can be replaced by organophilic clay in water.

A wide range of gelling materials are available from drilling fluids companies. Some are activated by time, temperature, or both; some by contact with downhole fluids, as those for gunk squeezes; some by deliberate mixing between two fluids pumped or squeezed separately down drillpipe and annulus; and some by shearing the fluid through the bit nozzles. All have regimes of downhole temperature, pressure, and chemistry where they work best, and should be chosen appropriately during planning.

Magzoub et al. (2020) give a comprehensive review of cross-linked polymer gels for treatment of lost circulation.

Cement Slurries. Cement is one of the oldest materials used in the oil and gas industry. It was certainly used by the Romans, who used a mixture of lime and pozzolan to create hydraulic cements, which could set under water.

Cement slurries have frequently been used as a lost circulation treatment. The selection of the cement type is dictated by the availability of the tested recipes and the products on the rig site. Sometimes, lost circulation material, such as glass or polyacrylamides fibers, are added to increase the success rate of stopping the losses (Taoutaou et al. 2006).

Cement can also be used as a discrete lost circulation treatment during drilling in the form of lost circulation plugs. Depending on the situation, the cement slurry is pumped through the BHA and drilling bit, or through open-ended drillpipe, when the risk of plugging the BHA is high.

Lightweight cement systems are most preferred to combat losses during primary cement jobs, because of the lower hydrostatic pressure and optimized rheological properties to control ECD during the placement of the fluids. These systems can consist of conventional extended cement, foamed cement, or optimized particle-distribution cement slurries.

Conventional extended cement uses extenders, such as glass bubbles; ceramic microspheres; liquid extenders, such as sodium silicate; or solid chemicals, such as bentonite, to replace part of the cement to control the density.

Foamed cement introduces gas—in this case nitrogen—into the cement matrix. Unlike the solid spheres, the nitrogen gas bubbles are compressible and, by expanding as pressure drops within a fracture or vug, play a big role in stopping lost circulation.

The optimized particle-distribution cements rely on the packing of the multiple particle sizes—usually three or four are used. The sizes—small, medium, large, and sometimes extra-large—vary from 6 to 1200 microns. The concentrations are chosen to control the density and rheology of the slurry, and the strength and permeability of the set cement (Veisi et al. 2015).

Particle Blends, Stress Caging, and Wellbore Strengthening. Since around 2000, there has been a lot of interest in using drilling fluids containing blends of sized particles aimed at blocking induced fractures and preventing their propagation. This was briefly described in Section 3.6 of Chapter 3 as a technique for drilling safely and economically through stacked reservoirs where the fracture gradient had been decreased by depletion.

One of the first approaches was the so-called *Stress Cage* technique, using hard particles (e.g., crushed marble) in a very low fluid-loss drilling fluid (Aston et al. 2004; Alberty and McLean 2004). Finite-element modeling was used to estimate the aperture of fractures likely to form under the wellbore pressures required by the drilling plan. Then particle sizes and concentrations were chosen to enter the fracture mouth and be trapped very close to the wellbore wall. This had two objectives. The first was to wedge open the mouth of the fracture and generate additional hoop stress (circumferential stress) around the wellbore (the Stress Cage), which was thought to inhibit further fracturing. The second was to establish a bridge of particles jammed in the fracture on which an impermeable filtercake could be deposited. The fluid behind the barrier could then leak off through the fracture faces, reducing the fluid pressure in the fracture to formation pressure and preventing its propagation. Treatments based on these concepts were highly successful for operators and allowed many otherwise difficult wells to be drilled.

A problem with marble in drilling fluids, however, is that it can be comminuted or ground up with passage through the mud pumps and nozzles, becoming too fine to block fractures effectively and also requiring constant additions to the mud system (followed by dilution). Another difficulty was that as the required well pressures increased and softer or less-stiff formations were targeted, the predicted fracture apertures and predicted marble particle sizes increased. These bigger particles, being relatively dense, could not be reliably suspended by the mud, and were also more vulnerable to comminution.

Another approach to the blocking of induced fractures was to use softer-sized particles, such as nut shells and graphitic carbon (for example, see van Oort et al. 2009). These were intended only to create a barrier within the fractures, and because they are compressible, they have negligible effect in raising the hoop stress. They are also less susceptible to comminution, so do not require frequent additions, and are more buoyant, which allows use of larger particles.

Fig. 5.8 shows the differences between the two approaches.

Choosing the right size range of bridging particles needs some input data. The width or aperture of the fracture at the wellbore determines whether or not a particle will be able to enter. The width is controlled largely by the Young's modulus (stiffness) of the rock, the length of the fracture (its extension away from the wellbore wall), the stress state in the formation, and the wellbore pressure. Young's modulus can be

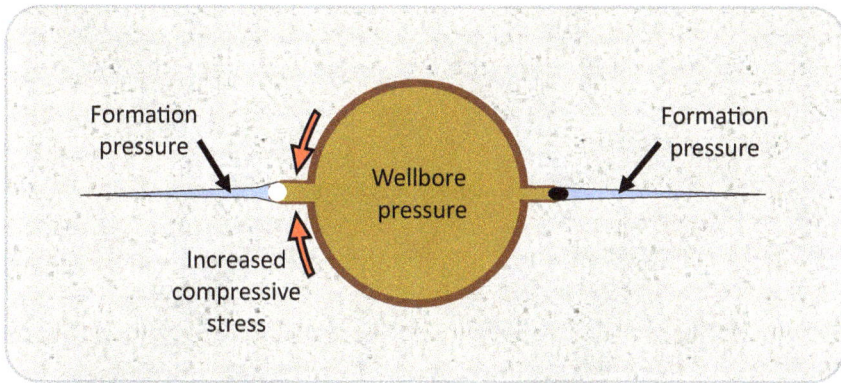

Fig. 5.8—Illustration of the differences between the Stress Cage idea (on the left) and the soft-particle approach (on the right). The Stress Cage method uses hard particles (e.g., marble), which can wedge open the fractures in the formation and consequently generate additional compressive hoop stress in the neighboring wellbore wall, potentially inhibiting further fractures. The soft particles used on the right are squeezed as the fracture closes on them and generate little additional stress. Both approaches, however, rely on generating a barrier to flow into the tip of the fracture. The pressure in the mouth end of the fracture is wellbore pressure, but in the tip end (blue), leakoff through the fracture faces means that the pressure drops to formation pressure, and this strongly inhibits further fracture growth.

obtained routinely from sonic logging data; smaller values of Young's modulus lead to wider fractures. Fracture length is difficult to calculate and is history-dependent, so most design approaches assume that the fracture length is 150 mm (6 in.) from wellbore wall to fracture tip. The formation stress state is today quite often known. There are simple models for it, usually based on Poisson's ratio (see Chapter 2), but it is a good idea to consult a geomechanics engineer. The wellbore pressure is the value required by the drilling plan to complete the casing interval in safety. Higher wellbore pressures lead to wider fractures, so high values of the required wellbore pressure, and/or low values of Young's modulus, mean larger particle sizes.

Choosing the right concentration of bridging particles uses the same data. If fracture width and length are known, the volume of the fracture per unit length of wellbore can be calculated. This volume of drilling fluid must contain enough particles to form a line of unit length; in other words, to form a bridge along unit length of wellbore. This is simple enough in principle, but rocks rarely behave in a simple way and so this calculated concentration should be used as a guide, rather than a definitive value. The Young's modulus of rocks varies from place to place, even within a single formation, so the width under a given wellbore pressure varies along the wellbore. Rock fracture surfaces are not smooth, so the aperture is not uniform, and the particles in the mud are not perfect spheres, so the criterion for trapping is not just diameter > aperture. The particle-trapping process is also not as simple as it seems at first sight; **Fig. 5.9** illustrates this. It is a snapshot from a video of a particle-laden fluid flowing in a slot whose aperture is larger than the particles at its entry, but

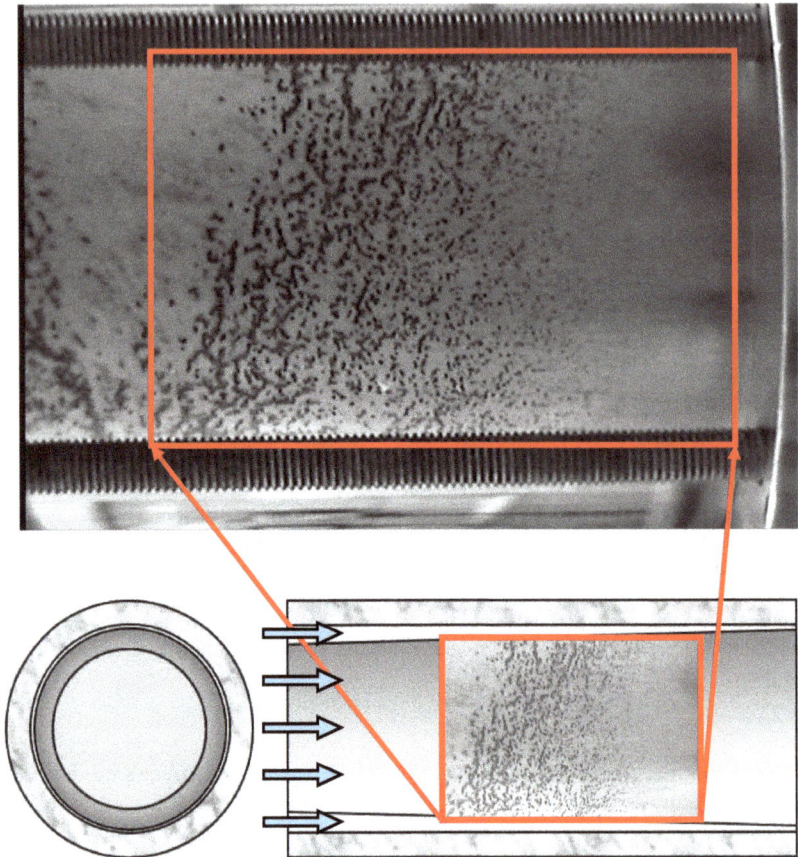

Fig. 5.9—Particle-trapping patterns in a tapering slot, acting as a fracture analogue. The slot is a cylindrical annulus between a clear outer pipe and a tapering aluminum inner mandrel. Particle-laden fluid is pumped into the left-hand end where the annulus is widest; it travels down the slot, and particles are trapped when the slot aperture becomes too small for them to pass. The fluid exits at the right. See text for discussion. (Image courtesy of P. Way.)

smaller at its exit (the slot is cylindrical, as shown in the accompanying diagram). This apparatus is also briefly discussed in Guo et al. (2014).

The particles in this case are graphitic carbon, transported in a polymer gel fluid. There is a very slight misalignment of the apparatus, so the slot at the top of the figure is wider than at the bottom; this means that particles can travel further down the slot at the top. Many particle bridges can be seen (coarser particles on the left, where the slot is wider), but these are incomplete. Small differences in particle size (within the specified particle-size distribution of the solid) lead to incomplete bridges at different slot widths. In a real fracture, this would allow fluid pressure to reach the tip and encourage propagation, even though the nominal particle concentration is enough to generate an effective barrier. The particles are also relatively weakly trapped, held in place by fluid drag; if the fluid flow changes, because a partial barrier is formed, for example, then the particle configuration can change. Elastic displacement of the fracture walls can also occur as fluid pressures change, and this may also disrupt particle bridges. To summarize, particle bridging is not as simple as it might seem because of small variations of fracture aperture and the intrinsic distribution of particle size. This means that much higher concentrations of particles may be needed than are given by the simple model previously mentioned.

The length and growth of the fracture is also not fully covered by the simple model. Fracture initiation is complex, especially a) in the wellbore geometry; b) with fluid pressure present, and c) for rocks that contain many pre-existing cracks and inhomogeneities. Fracture growth can be stable or unstable, depending on the stress state around the wellbore, the sizes of pre-existing flaws, and whether fluid pressure can penetrate them. This is beyond the scope of this book, but it is relevant to realize that fractures grow from either zero width and length or from the very small dimensions of a pre-existing flaw. These very small fractures can potentially be blocked at their mouths by very small particles. The significance of this will be discussed in the next section.

Finally, a new technology has emerged in the past few years that may change the game for fracture-blocking approaches. This is additive manufacturing or 3D printing. If the choice of particle shape and size is not restricted to bulk materials produced by grinding or chopping or other conventional industrial processes, could a shape be designed specifically to block fractures efficiently, and could this shape be produced economically by 3D printing? Although such materials are much more expensive than conventional lost circulation additives, they could be shaped to be much more effective, and so be required in far lower concentrations in the drilling fluid. There has been some work on this, but it is not yet in use. Way et al. (2016) discuss the types of shape that might be useful and some modifications to the current methods of additive manufacturing that improve the feasibility of economical production in bulk. Hitchcock (2020) shows how the idea could be used, even for large openings in the wellbore wall, to address total losses. As the costs of additive manufacturing come down, it may even be possible to have 3D printers at the wellsite, printing customized lost circulation materials on demand for particular diagnosed problems.

The Role of Filtercake. As the bit advances through a permeable formation, a fresh rock surface is exposed, and because the well is usually drilled overbalanced, wellbore fluid tends to flow into the rock. Provided that the pore size is not too large, some components of the drilling fluid will be filtered out onto the rock surface. These components might be, for example, drilled solids, barite, polymers, clays, emulsion droplets, or lost circulation materials. They form a mudcake or filtercake whose thickness and properties depend on the constituents, the pressure and flow regimes, temperature, and time. A typical filtercake formed from water-based mud might be 2 to 3 mm thick, with the consistency of toothpaste or putty (under atmospheric pressure). Oil-based mudcakes are typically thinner and weaker (Cerasi et al. 2001).

Filtercake can form very quickly; fluid-loss additives may be used in the drilling fluid to encourage this. It can also be washed away by flow from the bit nozzles or in the annulus. In the majority of cases, it should become thick enough and impermeable enough to slow fluid flow into the formation down to very low levels (if it doesn't, continuous loss into intact formations may occur, as described in Section 3.1 of Chapter 3).

The usual result, however, is that a layer of this filtercake is formed over the surface of the rock. If the rock surface on which the cake has formed is subjected to tensile stress, a fracture will eventually be initiated. If the cake is reasonably thick, it will be able to absorb the deformation imposed by the opening of the fracture, stretching a little to accommodate it. This means that it may remain intact over the opening of the fracture and prevent flow of wellbore fluid into the fracture, as illustrated in **Fig. 5.10.** Cook et al. (2016) describes this mechanism in more detail, including discussion of the various ways that filtercake can deform as it is stretched over an opening crack.

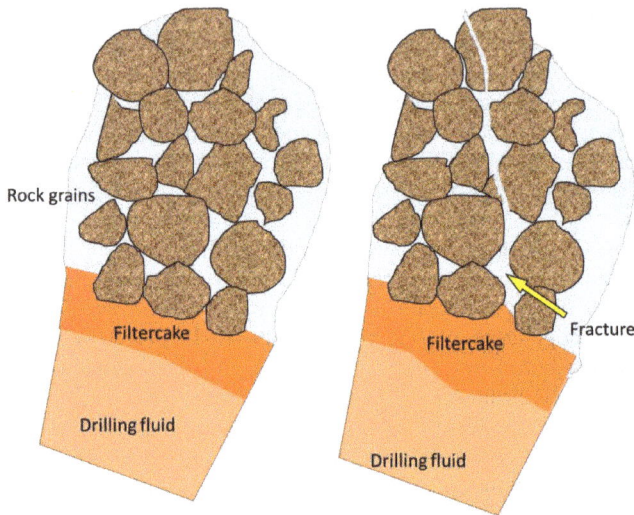

Fig. 5.10—Depiction of a few grains of the rock in the wellbore wall, illustrating the development of a filtercake and, on the right-hand side, the earliest stages of a fracture. Because the initial width of the fracture is comparable to the rock pore size, the filtercake does not invade it; because the filtercake is thick compared to the fracture opening, it stretches over the opening instead of breaking; adapted from Cook et al. (2016).

Cook et al. (2016) also describe experiments in fracture growth from a wellbore in which the filtercake thickness was varied. **Fig. 5.11** shows the equipment used: a biaxial loading rig acting on a cubical sandstone sample with a central hole representing the wellbore (Guo et al. 2014). By manipulating the wellbore pressure and external stresses, different thicknesses of filtercake could be grown on the inside of the wellbore, followed by pressurization to induce fracturing. **Fig. 5.12** shows an example in which a thicker filtercake allows a significantly higher wellbore pressure before fracture propagation occurs.

It might be thought from this that building a thick, strong filtercake is a good route to prevent lost circulation. It probably is, but it does come with some serious disadvantages for other aspects of well construction. Thick filtercake encourages differential sticking and may be hard to remove for cementing operations or

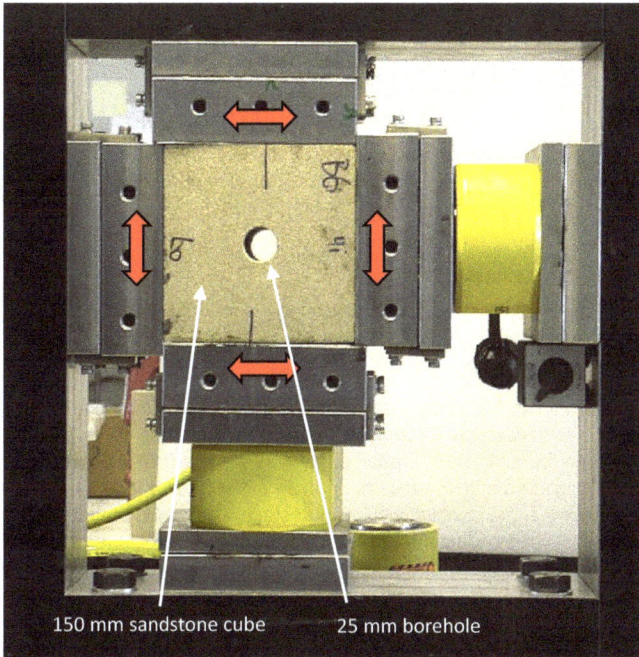

150 mm sandstone cube   25 mm borehole

Fig. 5.11—Biaxial loading system. This system was used for the fracturing experiments in Cook et al. (2016), Guo et al. (2014), and Therond et al. (2017). The sample is normally fully enclosed by steel plates at the front and back, with pressurized drilling fluids or cement slurry in the central wellbore. The red arrows indicate the allowed direction of movement of the inner parts of the loading platens relative to the outer parts; each platen can also overlap one of its neighbors, so the sample can be freely compressed.

production. A filtercake effective in strengthening the wellbore may nevertheless be built with additives in the drilling fluid. A US Patent Application in this area has been filed (Cook et al. 2020).

This clearer picture of the role of filtercake in fracture growth should be useful in understanding lost circulation phenomena and in designing new treatments.

Investigations of fracturing during cementing (Therond et al. 2017) by use of the equipment described, revealed that cement slurry also deposits filtercake on the wellbore wall and, in particular, at fracture openings. The high solids content of cement slurry means these cakes are thick, but in this situation, there is no need to worry about differential sticking.

In one experiment, the model wellbore (in Grinshill Sandstone) initially was filled with a nonaqueous drilling fluid and pressurized to form an initial filtercake. Then, the cement slurry described in **Table 5.1** was injected into the wellbore at a rate of 15 mL/min. The fluid injection pressure and the minimum stress as a function of time are shown in **Fig. 5.13**. The test rock fractured as the injection pressure increased, as indicated by the breaks in the pressure/time plots around 250 seconds. However,

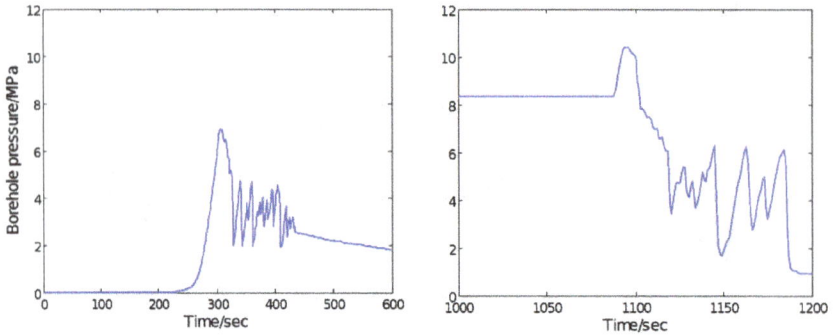

Fig. 5.12—Data from two fracturing experiments in the equipment of Fig. 5.11, using oil-based drilling fluid in sandstone. Each chart shows wellbore pressure vs. time as drilling fluid is pumped into the wellbore; the peak of the curve is interpreted as the point at which filtercake over a fracture mouth breaks, the fracture fills with mud, and then grows. The left-hand chart shows a test in which wellbore pressure was raised smoothly from zero, keeping the external stress constant. For the right-hand chart, wellbore pressure had previously been elevated to build a thicker filtercake, and the external stress had also been elevated to prevent premature fracture. From 1,090 seconds onward, the test conditions are the same on the right-hand chart as on the left-hand chart, but now when the wellbore pressure is increased at constant external stress, the wellbore has been strengthened by approximately 50%; from Cook et al. (2016).

Table 5.1—Characteristics of cement slurry used to fracture the sandstone sample (Therond et al. 2017).

Density	1965 kg/m³ (16.4 lbm/gal)
Solids content	43%
Plastic viscosity	155 cp
Yield stress	7.7 Pa (16 lbf/100 ft²)
Fluid loss (room temperature)	50 mL
Thickening time (77 F, 2000 psi)	10 hours

flow into the induced fractures was prevented by the slurry, and finally, a pressure of 16 MPa (2300 psi) was achieved. This is the maximum pressure allowed by the equipment; samples tested with drilling fluid alone typically fractured at approximately 9.7 MPa (1400 psi) pressure. Injection was stopped just after 400 seconds. When the end plates of the equipment were removed, a fracture was visible in the rock, with nodules of filtercake over the fracture entrance (Fig. 5.14). When removed from the test apparatus, the sample could be easily split into two pieces along the plane of the fracture. Minimum penetration of the cement slurry into the fracture was also observed (Fig. 5.15). Significant losses of fluid were prevented by the formation of the nodes of filtercake across the fracture entrance.

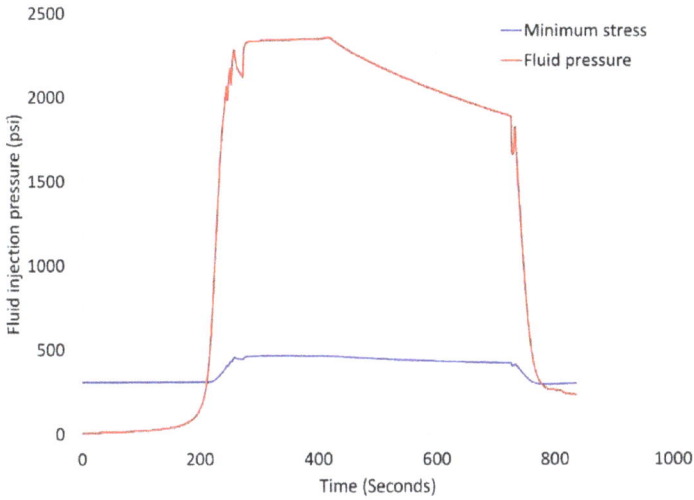

Fig. 5.13—Wellbore pressure and minimum stress vs. time for the cement slurry experiment described in the text (Therond et al. 2017).

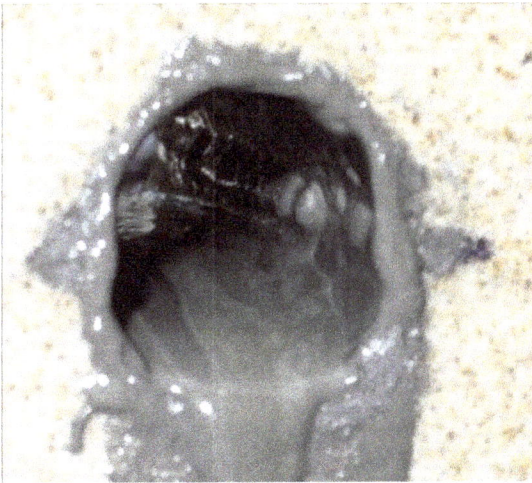

Fig. 5.14—Fracture visible from the injection end showing the cement nodules built up in the wellbore at the mouth of the horizontal fractures; after Therond et al. (2017).

The formation of cement slurry filtercake over the fracture mouth means that conventional cement slurries provide better lost circulation control than oil-based mud. For fractures with larger width, conventional cement slurries may require additional help (in this case the presence of fibers) to prevent losses.

Fig. 5.15—The sample split in two after the fracturing experiment. Note the grey filtercake on the surface of the wellbore, the buildups of filtercake at the fracture mouths, the limited penetration of slurry into the fracture on each side of the wellbore, and on the left-hand image, the additional small buildup of filtercake along a second fracture nucleating at right angles to the main fracture (this is linked to the increase in minimum stress seen in Fig. 5.14) (Therond et al. 2017).

Fluids Selection and Testing. Field Experience and Decision Trees: Most solutions initially are based on trial and error; once the losses can be relatively well-handled through this process, a lost circulation decision tree, as illustrated by Fig. 5.16, can be built for a specific field. The decision tree became a standard operating procedure

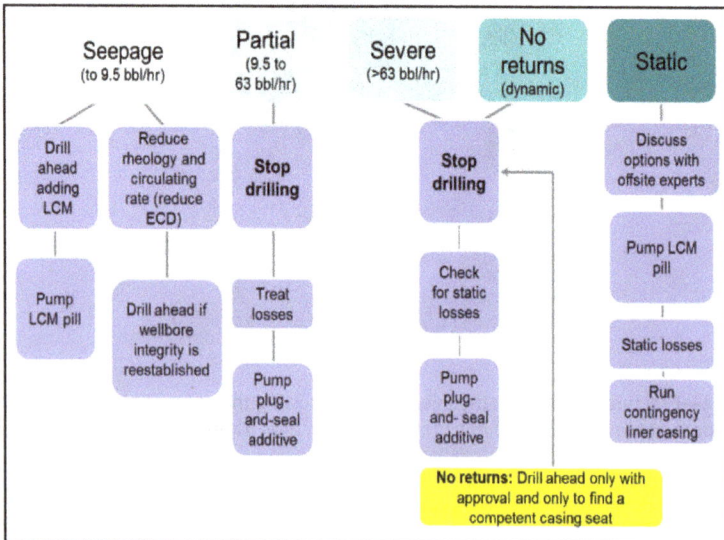

Fig. 5.16—A typical lost circulation decision tree is an iterative process based on the loss rate, guiding the user towards predefined lost circulation treatments.

and was used routinely to combat lost circulation and reduce operational and financial risks. The lost circulation decision tree is based on the severity of losses and generates sequential actions related to drilling tactics and drilling fluid changes.

The severity of losses is characterized by the loss rate: seepage, partial, severe, and total losses.

Seepage losses are less than 10 bbl/hr. They might occur in highly permeable rocks and in zones with fissures and microfractures. Partial losses vary from 10 to 200 bbl/hr and might also occur in highly permeable rocks and in zones with fissures and microfractures. Severe losses can reach 500 bbl/hr and can occur, for example, in highly fractured formations. Total losses are simply where there are no returns of the drilling fluid—this can be divided into total losses under dynamic conditions (i.e., while pumping) and static conditions.

Fluids Testing: With the advanced diagnostics and characterization models described in Chapter 4, treatments can be designed on the basis of the results of the modeling to meet the downhole conditions (for an example, see Mansour et al. 2018). To verify the solution proposed by the diagnostic model, a plugging efficiency test can be performed in the laboratory before each job, using a slot test. This reveals whether the chosen fluid can seal a model of the fracture in a metal disc, at the relevant temperature and under significant pressure.

The metal fracture model is selected on the basis of the diagnostic results, which generally estimate the fracture width. The slot size can range from 1 to 7 mm. The lost circulation pill is placed on top of the metal slot inside a high-pressure filtration cell. Nitrogen is then used to apply increasing pressure on the pill in increment of 100 psi (**Fig. 5.17**). Filtrate is collected below the cell, and its volume is measured. The test is finished when all the filtrate is collected, or the pressure drops, indicating breakdown of the plugging. The limitations of this test are that it can only be used at ambient temperature (unless a heating jacket is used), and the maximum pressure for the cell is 1000 psi (6.9 MPa).

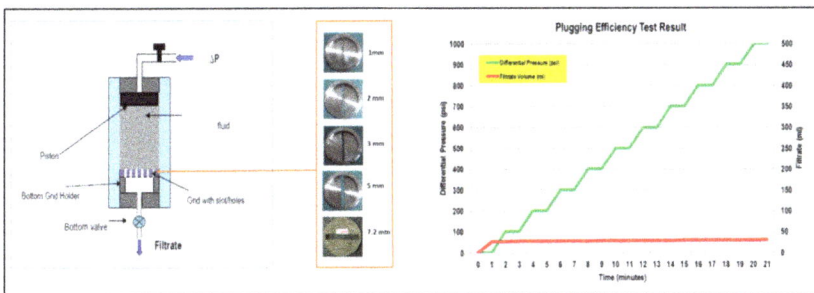

Fig. 5.17—Plugging test using a pressurized fluid loss cell and slots discs; from Taoutaou (2018).

To increase confidence in the effectiveness of the pill, an erodibility test can be performed by flowing fluid against the pill. This method is rarely used, however; it needs different equipment and setup and, of course, costs more.

Other test approaches use the permeability plugging apparatus (PPA). This equipment is used with drilling fluids in plugging tests (*ANSI/API RP 13B-1* 2009) and for testing the plugging capacity of drilling fluids (Davis et al. 1999). The PPA is used to determine the ability of particles in the drilling fluid to effectively bridge the pores in a filter medium. This equipment can be used with up to 2,000-psi pressure using nitrogen or carbon dioxide and with temperatures from ambient up to 500°F using a thermostatically controlled aluminum heating chamber.

The fluid cell is inverted with the pressure applied from the bottom of the cell, the filter medium on top, and the filtrate collected from the top. A small hydraulic hand-pump applies the cell pressure. The PPA (**Fig. 5.18**) can use porous ceramic or sintered metal disks with permeabilities varying from 100 md to 100 darcy, core samples, or beds of coated or uncoated sand, as well as machined stainless-steel discs.

These testing methods provide the user with a greatly improved picture of what is happening downhole.

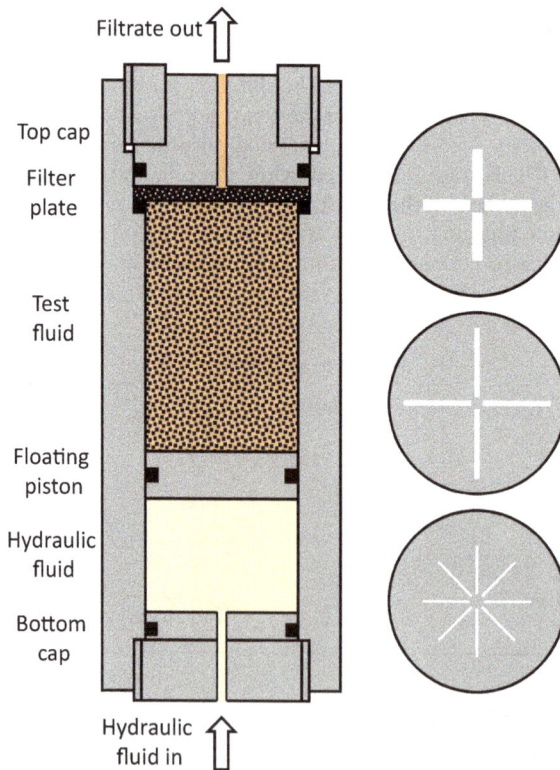

Fig. 5.18—Depiction of PPA, for use at high pressure and temperature. The filter plate can be either a ceramic disc with predefined permeability and pore size, or discs with slots having different lengths and widths (to represent fracture openings), as shown on the right-hand side.

5.3. Real-Time Techniques During Drilling, Casing, and Cementing.

5.3.1. Drilling. There are many ways in which drilling operations can improve or worsen lost circulation problems. Some of these need advanced planning, others involve actions while drilling.

The aim can be

- To keep the ECD at all times below the fracture gradient (established with the help of a geomechanics engineer and local experience) everywhere in the open hole. This would be the objective if, for example, the mud weight window was relatively wide
- To maintain the hole conditions (such as plugs of lost circulation material or wellbore-strengthening particles) that allow the ECD to exceed the fracture gradient without severe losses. This would be the objective if, for example, the mud weight window was very narrow

The ECD can be reduced with some standard procedures:

- Ensuring the drilling fluid is in good condition, with its rheology and density within specification
- Ensuring the drillstring is rotated to break the gel before moving down, if the fluid has been stationary and has had time to gel
- Limiting rate of penetration so that the fluid in the annulus does not become too dense with cuttings
- Choosing bit, BHA, and drillpipe diameters to maximize the flow area in the annulus
- Carefully cleaning the hole to keep the annulus clear, especially if wellbore instability is generating cavings
- Reducing surge pressures by limiting tripping-in speed and the speed of all downward movements of the drillstring.

These are straightforward procedures, but it is important to know whether they are working. There are many tools available from service companies that measure annulus pressure and transmit the value to surface. One of these should be in the drillstring and constantly monitored whenever close control of the ECD is required. Some can operate in record mode so that pressure excursions during tripping and connections, among other things, can be monitored, models for wellbore hydraulics can be calibrated, and rig procedures modified accordingly.

If the ECD is above the fracture gradient, using lost circulation materials or similar approaches, then there should be a focus on keeping the fracture-blocking agents in place. Swabbing should be avoided, so close control is needed of drillstring speed while tripping out. Hole-cleaning procedures should be gentle, without back reaming, to prevent the buildup of cuttings beds, which can be plowed by the BHA during trips, leading to packoffs and sudden ECD increase below the bit. Aggressive bits with strong side-cutting ability should not be used, nor close-fitting stabilizers with sharp edges. Any vibration or rough drilling tends to scrape the blocking material away from the fracture mouths.

5.3.2. Running Casing. It is important that everyone involved realizes that the section is not safely completed when the drill bit begins its trip out after reaching section total depth. Subsequent operations play major roles in the overall success of the well and must also be carefully designed, executed, and monitored. Running of the casing in the hole and working it up and down when encountering tight spots, for example, produces downhole pressure variations and the displacement of the drilling fluids in the annulus. The pressure changes are governed by the geometry of the casing and hole, the velocity of displacement, and the rheology of the annular fluid. These pressure changes add to the original hydrostatic pressure of the drilling fluid to produce surge or swab pressures, just as during drilling operations.

When the surge pressure is higher than the fracture gradient, it can cause induced fractures or open up existing fractures, leading to loss of drilling fluids into the formation and all the accompanying problems.

Controlling the speed of casing when running in hole is very important to allow the casing to reach the total depth without causing lost circulation. Best practices and operating procedures should be put in place to control this operation. The first action is to establish boundaries and to determine the maximum casing running speeds.

The maximum-allowed wellbore pressure should already have been established before drilling the well, using either geomechanics models or local experience or, ideally, both. The location of the point in the open hole at which the formation is most susceptible to fracturing is often assumed to be at the previous casing seat, but this is not necessarily the case. Changes in rock properties, natural fractures, and fractures induced by the drilling operations may all occur at greater depths.

After determining the maximum-allowable pressure limit, hydraulics modeling is required to determine the maximum running speed of the casing that does not exceed the limit. Calculations of this kind have been available for many years, with varying degrees of sophistication (e.g., Chukwu 1995). If a caliper log of the open hole can be obtained from logging while drilling during the final trip out of the hole, the measured hole size can be used to refine these estimates.

Orellan et al. (2010) describe a comprehensive approach to mitigate severe lost circulation problems when running casing. The primary methods are to control the bypass area between casing strings, to control the speed of the casing being run in hole, to control the viscosity of the drilling fluids, and to use an auto-fill system that allows fluid to flow freely into the casing so that the casing is not pushing excessive volumes of fluids up the annulus.

5.3.3. Cementing. Real-time software models have been developed to monitor the primary cement jobs, in general, and lost circulation, in particular, by looking at the variation of real-time ECD (Contreras et al. 2017). The real-time monitoring software provides a solution to the lack of a direct flow rate, computing and displaying the return rate using paddle meter measurements and the derivative over time of the returned volume observed in the rig tanks. The advantage it offers is that in conditions in which losses are present and the loss-zone interval can be estimated, a decision to reduce the pump rate to control ECD can take place on the spot and a simulation run in real time. If a lost circulation treatment is needed, it can be placed more accurately because the real-time hydraulic simulator can calculate the rate of losses going into the defined loss zone at each timestep, and the fluids positioning in the annulus can be recomputed.

Figs 5.19 and 5.20 show an example of a lost circulation event during cementing using real-time returns monitoring. Measurements showed a volume loss of

1006 bbl, and no lift pressure was detected. Unfortunately, in this case, no mitigation was possible, except the reduction of the pumping rate to reduce the ECD.

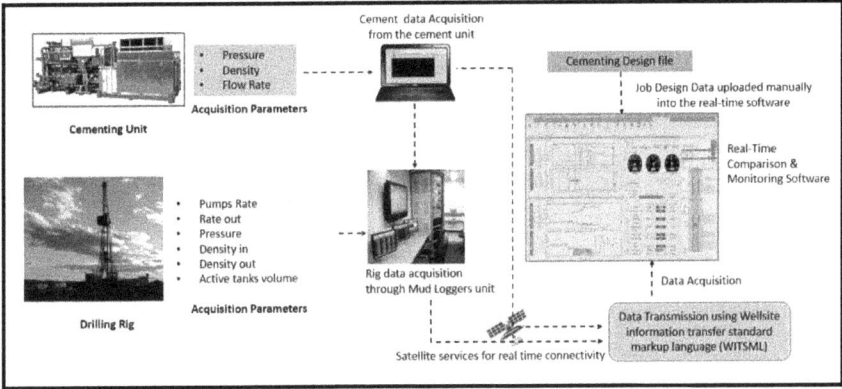

Fig. 5.19—Principles of real-time monitoring and analysis of cementing operations; after Contreras et al. (2017).

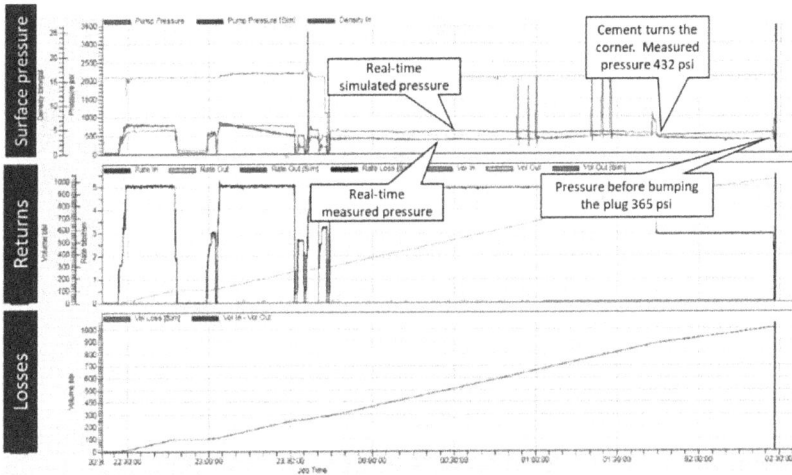

Fig. 5.20—Real-time monitoring and analysis of cementing operations, including lost circulation events. The software also allows real-time display of the fluid interface positions in the pipe and annulus. In this case, a volume loss of 1006 bbl and no lift pressure were detected; after Contreras et al. (2017).

5.3.4. Placement of Lost Circulation Treatments. There are several ways of spotting a lost circulation treatment, as mentioned in Section 5.2.4: through the BHA and drill bit, through open-ended drillpipe after the BHA is pulled out, or through a diverter sub. The latter allows pumping from the drillpipe directly into the annulus without flowing through BHA and bit—there are several tools available from service companies. A range of activation mechanisms are used; they can be activated by dropping a ball, sending pressure pulses, or radio-frequency identification (RFID) technology, as described by Valverde et al. (2016).

For the RFID systems, the circulating valve is activated using a small robust RFID tag, which is dropped at the surface into the drillpipe. A built-in antenna in the diverter sub detects the tag and activates an electric motor driving a hydraulic pump that opens or closes the side-exit valve. The opening and closing can be performed as many times as required, which is more difficult with a ball-drop system. In the event of failure of the antenna, the sub can be activated using a dart.

When pumping through the drilling assembly (BHA and drill bit) and open-ended drillpipe, the following considerations must be taken into account:

- A thorough risk assessment must be performed to establish
  - The minimum opening through which the lost circulation pill will be pumped
  - The presence of measurement tools such as MWD/logging while drilling, which can be harmed by the lost circulation pill
  - Drillpipe screens/strainers that could restrict the lost circulation pills
  - The total flow area of the bit nozzles to assess any premature plugging of the drill bit, especially when the lost circulation pill is particle- and fiber-laden fluid
- Pumping the lost circulation pill
  - When pumping the lost circulation pill through an open-ended drillpipe, the pill will be placed using a "balanced plug" technique; in this case, thorough calculations are performed to balance the fluids during the pumping and while pulling the pipe out of the pill after the placement to avoid contamination of the fluids, which can jeopardize the effectiveness of the lost circulation pill
  - Consider the use of the diverter sub to avoid the jetting effect of the fluids, which cause turbulence and contamination of the fluids
  - When pumping through the BHA and drill bit, the assembly should be placed in front of the lost circulation zone; the pill is pumped through to finally exit the drill bit nozzles
  - In all cases, the pumping schedule is carefully chosen to avoid any surge of the well and to avoid fluids contamination

Summary

- *In the well planning phase, consider drilling engineering solutions to lost circulation risks, such as managed pressure drilling and cementing, casing while drilling, and expandable tubulars.*
- *Assemble as much information as possible from offset wells about depths and rates of lost circulation events, any actions that triggered or preceded them, and any procedures and materials used to reduce or eliminate losses, successful and unsuccessful.*
- *Consult a geomechanics engineer to get an estimate of the minimum stress and how it varies along the well path.*

- *Ensure that everyone involved, including the rig and casing crews, understands the importance of avoiding swab and surge. Set limits for the maximum speed of movement of the drillstring and casing and stick to them. While running the casing, use automated fill-up tools to reduce ECD variations.*
- *If the chosen approach to mitigating lost circulation is to block or bridge fractures, be aware of the impact of drilling hardware and vibrations on the integrity of the bridges.*
- *Prevent lost circulation during cementing by designing cementing fluids with controlled densities and viscosities to reduce the friction pressure in the annulus and control ECD. Use the software tools now available to help with this complex process and to validate the design.*
- *Use data from surface and downhole in a diagnostic model to characterize the losses and learn as much as possible about their mechanisms. When diagnostics models are not available, this can be done using a lost circulation decision tree.*
- *Use real-time data—especially annular pressure while drilling—and real-time models to monitor drilling and cementing and to prevent and/or mitigate losses.*

5.4. References.

Alberty, M. W. and McLean, M. R. 2004. A Physical Model for Stress Cages. Paper presented at the SPE Annual Technical Conference and Exhibition, Houston, Texas, 26–29 September. SPE-90493-MS. https://doi.org/10.2118/90493-MS.

ANSI/API RP 13-B, Recommended Practice for Field Testing Water-Based Drilling Fluids, fourth edition, 2009. Washington DC: API.

Aston, M. S., Alberty, M. W., McLean, M. R. et al. 2004. Drilling Fluids for Wellbore Strengthening. Paper presented at the IADC/SPE Drilling Conference, Dallas, Texas, 2–4 March. SPE-87130-MS. https://doi.org/10.2118/87130-MS.

Bjorkevoll, K. S., Molde, D. O., Rommetveit, R. et al. 2008. MPD Operation Solved Drilling Challenges in a Severely Depleted HP/HT Reservoir. Paper presented at the IADC/SPE Drilling Conference, Orlando, Florida, USA, 4–6 March. SPE-112739-MS. https://doi.org/10.2118/112739-MS.

Cales, G. L. 2003. The Development and Applications of Solid Expandable Tubular Technology.Paper presented at the Canadian International Petroleum Conference, Calgary, Alberta, 10–12 June. PETSOC-2003-136. https://doi.org/10.2118/2003-136.

Cerasi, P., Ladva, H. K., Bradbury, A. J. et al. 2001. Measurement of the Mechanical Properties of Filtercakes. Paper presented at the SPE European Formation Damage Conference, The Hague, The Netherlands, 21–22 May. SPE-68948-MS. https://doi.org/10.2118/68948-MS.

Chapman, M. T. 1913. Well-Sinking Apparatus. US Patent No. 1079539A.

Chukwu, G. A. 1995. A Practical Approach for Predicting Casing Running Speed From Couette Flow of Non-Newtonian Power-Law Fluids. Paper presented at the SPE Western Regional Meeting, Bakersfield, California, 8–10 March. SPE-29638-MS. https://doi.org/10.2118/29638-MS.

Contreras, J., Bogaerts, M., Griffin, D. et al. 2017. Real-Time Monitoring and Diagnoses on Deepwater Cement Barrier Placement: Case Studies from the Gulf of Mexico and Atlantic Canada. Paper presented at the Offshore

Technology Conference, Houston, Texas, 1–4 May. OTC-27797-MS. https://doi.org/10.4043/27797-MS.

Cook, J., Guo, Q., Way, P. et al. 2016. The Role of Filtercake in Wellbore Strengthening. Paper presented at the IADC/SPE Drilling Conference and Exhibition, Fort Worth, Texas, 1–3 March. SPE-178799-MS. https://doi.org/10.2118/178799-MS.

Cook, J. M., Guo, Q., Way, P. W. et al. 2020. Method and Model for Wellbore Strengthening By Filtercake. US Patent No. 2020/0284109 A1 (10 September 2020).

Davis, N., Mihalik, P., Lundie, P. R. et al. 1999. New Permeability Plugging Apparatus Procedure Addresses Safety and Technology Issues. Paper presented at the SPE/IADC Drilling Conference, Amsterdam, The Netherlands, 9–11 March. SPE-52815-MS. https://doi.org/10.2118/52815-MS.

Dipura, Y. S., Ardiyaprana, F. B., and Putra, E. M. 2018. Pressurized Mud Cap Drilling Drastically Improves Drilling Efficiency in Exploration Well, South Sumatra. Paper presented at the SPE/IADC Middle East Drilling Technology Conference and Exhibition, Abu Dhabi, UAE, 29–31 January. SPE-189398-MS. https://doi.org/10.2118/189398-MS.

Filippov, A., Mack, R., Cook, L. et al. 1999. Expandable Tubular Solutions. Paper presented at the SPE Annual Technical Conference and Exhibition, Houston, Texas, 3–6 October. SPE-56500-MS. https://doi.org/10.2118/56500-MS.

Frink, P. 2006. Managed Pressure Drilling—What's in a Name? *Drilling Contractor* (March/April): 36–39. https://iadc.org/dcpi/dc-marapr06/Mar06-suri.pdf.

Guo, Q., Cook, J., Way, P. et al. 2014. A Comprehensive Experimental Study on Wellbore Strengthening. Paper presented at the IADC/SPE Drilling Conference and Exhibition, Fort Worth, Texas, 4–6 March. SPE-167957-MS. https://doi.org/10.2118/167957-MS.

Fontenot, K., Highnote, J., Warren, T. et al. 2003. Casing Drilling Activity Expands in South Texas. Paper presented at the SPE/IADC Drilling Conference, Amsterdam, The Netherlands, 19–21 February. SPE-79862-MS. https://doi.org/10.2118/79862-MS.

Hitchcock, G. 2020. Additive Manufactured Shapes Used to Cure Total Lost Circulation Events. Paper presented at the Offshore Technology Conference, Houston, Texas, 4–7 May. OTC-30757-MS. https://doi.org/10.4043/30757-MS.

Houtchens, B. D., Warren, T. M., Tessari, R. M. et al. 2007. Applying Risk Analysis to Casing While Drilling. Paper presented at the SPE/IADC Drilling Conference, Amsterdam, The Netherlands, 20–22 February. SPE-105678-MS. https://doi.org/10.2118/105678-MS.

Lavrov, A. 2017. Lost Circulation in Primary Well Cementing, *Energy Procedia*, vol. 114 (July): 5,182–5,192. https://doi.org/10.1016/j.egypro.2017.03.1672.

Pasteris, M., Taoutaou, S., Simanjuntak, A. et al. 2014. Reinforced Composite Mat, a Proven Solution for Lost Circulation Control in Sumatra Field. Paper presented at the SPE Asia Pacific Oil & Gas Conference and Exhibition, Adelaide, Australia, 14–16 October. SPE-171412-MS. https://doi.org/10.2118/171412-MS.

Magzoub, M. I., Salehi, S., Hussein, I.A. et al. (2020) Loss circulation in drilling and well construction: The significance of applications of crosslinked polymers in wellbore strengthening: A review. *J. Pet. Sci. Eng.* **185** (February). 106653. https://doi.org/10.1016/j.petrol.2019.106653.

Mansour, A. K. and Taleghani, A. D. 2018. Smart Loss Circulation Materials for Drilling Highly Fractured Zones. Paper presented at the SPE/IADC Middle East

Drilling Technology Conference and Exhibition, Abu Dhabi, UAE, 29–31 January. SPE-189413-MS. https://doi.org/10.2118/189413-MS.

Orellan, S., May, R., Bedino, H. et al. 2010. Design of "Anti Surge" Methodology to Mitigate Severe Lost Circulation While Running Non-Conventional Casing/Liner Sizes to Isolate Salt and Clay Domes in Deep Wells in Mexico South. Paper presented at the IADC/SPE Asia Pacific Drilling Technology Conference and Exhibition, Ho Chi Minh City, Vietnam, 1–3 November. SPE-135905-MS. https://doi.org/10.2118/135905-MS.

Rajabi, M. M., Rohde, B., Maguire, N. et al. 2012. Successful Implementations of Tophole Managed Pressure Cementing and Managed Pressure Drilling in the Caspian Sea. Paper presented at the SPE/IADC Managed Pressure Drilling and Underbalanced Operations Conference and Exhibition, Milan, Italy, 20–21 March. SPE-156889-MS. https://doi.org/10.2118/156889-MS.

Riddoch, J., Wuest, C., and Toralde, J. S. S. 2016. Managing Constant Bottom Hole Pressure with Continuous Flow Systems. OTC-26752-MS. Paper presented at the Offshore Technology Conference Asia, Kuala Lumpur, Malaysia, 22–25 March. OTC-26752-MS. https://doi.org/10.4043/26752-MS.

Sánchez, F., Houqani, S., Turki, M. et al. 2012. Casing While Drilling (CwD): A New Approach to Drilling Fiqa Formation in The Sultanate of Oman—A Success Story. SPE Drill & Compl 27 (02): 223–232. SPE-136107-PA. https://doi.org/10.2118/136107-PA.

Sanders, M. W, and Houston, K. A. 2014. Successful Planning and Integration of a HFHS Lost Circulation Treatment for Challenging North Sea Brownfield Development. Paper presented at the International Petroleum Technology Conference, Kuala Lumpur, Malaysia, 10–12 December. IPTC-17717-MS. https://doi.org/10.2523/IPTC-17717-MS.

Sanders, M. W., Scorsone, J. T., and Friedheim, J. E. 2010. High-Fluid-Loss, High-Strength Lost Circulation Treatments. Paper presented at the SPE Deepwater Drilling and Completions Conference, Galveston, Texas, 5–6 October. SPE-135472-MS. https://doi.org/10.2118/135472-MS.

Siddiqi, F. A., Riskiawan, A., Al-Yami, A. et al. 2016. Successful Managed Pressure Cementing With Hydraulic Simulations Verification in a Narrow Pore-Frac Pressure Window Using Managed Pressure Drilling in Saudi Arabia. Paper presented at the SPE Annual Technical Conference and Exhibition, Dubai, UAE, 26–28 September. SPE-182500-MS. https://doi.org/10.2118/182500-MS.

Stave, R. 2014. Implementation of Dual Gradient Drilling. Paper presented at the Offshore Technology Conference, Houston, Texas, 5–8 May. OTC-25222-MS. https://doi.org/10.4043/25222-MS.

Taoutaou, S. 2018. Lost Circulation—A Challenge We Must Address. Presented as an SPE Distinguished Lecture during the 2017–2018 season; March 2018 lecture presented at Duliajan, India.

Taoutaou, S., Ashraf, S., Abdel Hadi, K. 2013. Compositions and Methods for Servicing Subterranean Wells. US Patent No. US20130043026A1 (21 February 2013).

Taoutaou, S., Shuttleworth, N., VanderPlas, K. et al. 2006. An Innovative Inert Material To Cure the Losses in the Brent Depleted Reservoirs—North Sea Case Histories. Paper presented at the Abu Dhabi International Petroleum Exhibition and Conference, Abu Dhabi, UAE, 5–8 November. SPE-100934-MS. https://doi.org/10.2118/100934-MS.

Therond, E., Taoutaou, S., James, S. G. et al. 2017. Understanding Lost Circulation While Cementing: Field Study and Laboratory Research. Paper presented at the SPE/IADC Drilling Conference and Exhibition, The Hague, The Netherlands, 14–16 March. SPE-184673-MS. https://doi.org/10.2118/184673-MS.

Valverde, E. and Goodwin, A. 2016. Radio Frequency Identification (RFID)-Enabled Circulation Sub Precisely Spots Loss Circulation Material in Critical Interval. Paper presented at the Offshore Technology Conference, Houston, Texas, 2–5 May. OTC-26970-MS. https://doi.org/10.4043/26970-MS.

van Oort, E., Friedheim, J. E., Pierce, T. et al. 2009. Avoiding Losses in Depleted and Weak Zones by Constantly Strengthening Wellbores. Paper presented at the SPE Annual Technical Conference and Exhibition, New Orleans, Louisiana, 4–7 October. SPE-125093-MS. https://doi.org/10.2118/125093-MS.

Veisi, M. S., Taoutaou, S., Steven, A. et al. 2015. Engineered Highly Crush-Resistant Cement Slurry to Prevent Lost Circulation. Paper presented at the SPE/IATMI Asia Pacific Oil & Gas Conference and Exhibition, Nusa Dua, Bali, Indonesia, October. SPE-176038-MS. https://doi.org/10.2118/176038-MS.

Wahid, F., Tajalie, A. F. A., Taoutaou, S. et al. 2014. Successful Cementing of Ultra HTHP Wells Under Managed Pressure Drilling Technique. Paper presented at the International Petroleum Technology Conference, Kuala Lumpur, Malaysia, 10–12 December. IPTC-17739-MS. https://doi.org/10.2523/IPTC-17749-MS.

Way, P., Snoswell, D., Cook, J. M. et al. 2016. Solids in Borehole Fluids. US Patent No. 20160244654 A1.

www.ingramcontent.com/pod-product-compliance
Lightning Source LLC
Chambersburg PA
CBHW060322220326

41598CB00027B/4400